Hanns Soblik

Essential proteins of parasitic nematodes:

AF092502

Hanns Soblik

Essential proteins of parasitic nematodes:

Proteomic analysis of excretory/secretory proteins from parasitic and free-living stages of Strongyloides ratti

Südwestdeutscher Verlag für Hochschulschriften

Impressum/Imprint (nur für Deutschland/ only for Germany)
Bibliografische Information der Deutschen Nationalbibliothek: Die Deutsche Nationalbibliothek verzeichnet diese Publikation in der Deutschen Nationalbibliografie; detaillierte bibliografische Daten sind im Internet über http://dnb.d-nb.de abrufbar.
Alle in diesem Buch genannten Marken und Produktnamen unterliegen warenzeichen-, marken- oder patentrechtlichem Schutz bzw. sind Warenzeichen oder eingetragene Warenzeichen der jeweiligen Inhaber. Die Wiedergabe von Marken, Produktnamen, Gebrauchsnamen, Handelsnamen, Warenbezeichnungen u.s.w. in diesem Werk berechtigt auch ohne besondere Kennzeichnung nicht zu der Annahme, dass solche Namen im Sinne der Warenzeichen- und Markenschutzgesetzgebung als frei zu betrachten wären und daher von jedermann benutzt werden dürften.

Verlag: Südwestdeutscher Verlag für Hochschulschriften Aktiengesellschaft & Co. KG
Dudweiler Landstr. 99, 66123 Saarbrücken, Deutschland
Telefon +49 681 37 20 271-1, Telefax +49 681 37 20 271-0, Email: info@svh-verlag.de
Zugl.: Hamburg, Universität Hamburg, Diss., 2009

Herstellung in Deutschland:
Schaltungsdienst Lange o.H.G., Berlin
Books on Demand GmbH, Norderstedt
Reha GmbH, Saarbrücken
Amazon Distribution GmbH, Leipzig
ISBN: 978-3-8381-1024-0

Imprint (only for USA, GB)
Bibliographic information published by the Deutsche Nationalbibliothek: The Deutsche Nationalbibliothek lists this publication in the Deutsche Nationalbibliografie; detailed bibliographic data are available in the Internet at http://dnb.d-nb.de.
Any brand names and product names mentioned in this book are subject to trademark, brand or patent protection and are trademarks or registered trademarks of their respective holders. The use of brand names, product names, common names, trade names, product descriptions etc. even without a particular marking in this works is in no way to be construed to mean that such names may be regarded as unrestricted in respect of trademark and brand protection legislation and could thus be used by anyone.

Publisher:
Südwestdeutscher Verlag für Hochschulschriften Aktiengesellschaft & Co. KG
Dudweiler Landstr. 99, 66123 Saarbrücken, Germany
Phone +49 681 37 20 271-1, Fax +49 681 37 20 271-0, Email: info@svh-verlag.de

Copyright © 2009 by the author and Südwestdeutscher Verlag für Hochschulschriften Aktiengesellschaft & Co. KG and licensors
All rights reserved. Saarbrücken 2009

Printed in the U.S.A.
Printed in the U.K. by (see last page)
ISBN: 978-3-8381-1024-0

Proteomic analysis of excretory/secretory proteins from parasitic and free-living stages of *Strongyloides ratti*

Dissertation

zur Erlangung des Doktorgrades des Department Chemie der Universität Hamburg

vorgelegt von

Hanns Soblik

aus Fahrdorf

Hamburg 2009

The present thesis was carried out between January 2006 and Febuary 2009 at the Bernhard Nocht Institute for Tropical Medicine, Hamburg, Germany and at the Proteomics Center at Children's Hospital, Boston, USA.

1. Reviewer: PD Dr. Norbert Brattig
2. Reviewer: Prof. Dr. Peter Heisig

Disputation date: 10th of July 2009

Für meine Eltern

Table of Contents

1 Introduction 1
1.1 The life cycle of *Strongyloides* spp. 1
1.2 Pathology of *Strongyloides* infections 3
1.3 Treatment of strongyloidiasis 4
1.4 Diagnosis of strongyloidiasis 5
1.5 Immune response to *Strongyloides* 6
1.6 *Strongyloides ratti* as a laboratory model 7
1.7 Protein secretion of nematodes 8
1.8 Objectives 11

2 Animals, Materials and Methods 12
2.1 Animals 12
2.2 Materials 12
    2.2.1 Devices 12
    2.2.2 Kits 13
    2.2.3 Solutions 14
    2.2.4 Culture Media and supplements 16
    2.2.5 Plasmids 17
    2.2.6 Oligonucleotides 17
    2.2.7 Enzymes 18
2.3 Methods 19
    2.3.1 Working with *Strongyloides ratti* 19
        2.3.1.1 Infection of hosts 19
        2.3.1.2 Charcoal culture and Baermann apparatus 19
        2.3.1.3 Preparation of iL3 20
        2.3.1.4 Preparation of the free-living stages 21
        2.3.1.5 Preparation of parasitic females 21
        2.3.1.6 Preparation of E/S products 21
        2.3.1.7 Preparation of worm extracts 22
        2.3.1.8 Whole worm analysis 22

2.3.2 Molecular biological methods 23

    2.3.2.1 *E. coli* culture 23

    2.3.2.2 Generation of competent bacteria and transformation 23

    2.3.2.3 Plasmid preparations 23

    2.3.2.4 Expression of *S. ratti* galectin-3 in *E. coli* 24

    2.3.2.5 Purification of recombinant proteins by affinity chromatography 24

    2.3.2.6 Total RNA isolation 25

    2.3.2.7 Reverse transcription 25

    2.3.2.8 Polymerase chain reaction 26

    2.3.2.9 Purification of DNA fragments 26

    2.3.2.10 5'- and 3'-cDNAs amplification 27

    2.3.2.11 Agarose gel electrophoresis 27

    2.3.2.12 Determination of nucleic acid concentrations 27

    2.3.2.13 Ligating of DNA 27

    2.3.2.14 Restriction analysis 28

    2.3.2.15 DNA sequencing 29

2.3.3 Biochemical methods 29

    2.3.3.1 Determination of protein concentration by Bradford assay 29

    2.3.3.2 SDS-polyacrylamide gel electrophoresis (PAGE) 29

    2.3.3.3 Coomassie staining of polyacrylamide gels 30

    2.3.3.4 Silver staining of polyacrylamide gels 30

    2.3.3.5 Substrate gel electrophoresis - zymogram 30

    2.3.3.6 Gelatin gel overlay 30

    2.3.3.7 Lactose affinity separation 31

    2.3.3.8 One dimensional-electrophoresis and band excision 31

    2.3.3.9 Tryptic digestion 32

    2.3.3.10 Liquid chromatography - tandem mass spectrometry (LC-MS/MS) 33

Table of Contents

2.3.4 Bioinformatic procedures 38
    2.3.4.1 Database searches 38
    2.3.4.2 Phylogenetic analysis 39
2.3.5 Immunological tests 39
    2.3.5.1 Western Blot analysis 39
    2.3.5.2 ELISA (Enzyme-linked immuno-sorbent assay) 39

3 Results 41

3.1 Establishing the *S. ratti* life cycle 41
3.2 Optimising and processing of E/S products 45
    3.2.1 Dependence on the number of larvae 45
    3.2.2 Detection of the metalloprotease 46
    3.2.3 Size determination of the metalloprotease 46
    3.2.4 Dependence on the incubation times 47
    3.2.5 Dependence on the incubation temperature 48
    3.2.6 Inhibition of protein synthesis 49
    3.2.7 Differences in protein secretions and crude extracts 50
    3.2.8 Differences in protein secretion among various stages 51
    3.2.9 Antibody recognition of E/S products 52
3.3 Mass spectrometry 53
    3.3.1 Comparison of proteins secreted from different stages 54
    3.3.2 Abundant proteins in E/S products shared by all stages 57
    3.3.3 Stage-related proteins 61
        3.3.3.1 Proteins enriched in infective larvae 61
        3.3.3.2 Proteins enriched in parasitic females 63
        3.3.3.3 Proteins enriched in the free-living stages 65
    3.3.4 Demonstration of differentially expressed proteins applying PCR analysis 66
3.4 Selected candidate functional proteins 67

	3.4.1	Identification and analysis of *S. ratti* galectins	67
		3.4.1.1 Completion of galectin sequences	70
		3.4.1.2 Sequence analysis of galectins	71
		3.4.1.3 Phylogenetic analysis of galectins	74
		3.4.1.4 Isolation of native galectins	75
		3.4.1.5 Prokaryotic expression of *Sr*-Gal-3	77
		3.4.1.6 Antibody recognition of *Sr*-Gal-3	77
		3.4.1.7 Sugar-binding assay of galectins	78
	3.4.2	Identification and analysis of a *S. ratti* prolyl oligopeptidase	81
		3.4.2.1 Completion of the *Sr*-POP-1 sequence	81
		3.4.2.2 Mass spectrometric analysis of *Sr*-POP-1	83
		3.4.2.3 Sequence analysis of *Sr*-POP-1	84
		3.4.2.4 Phylogenetic analysis of *Sr*-POP-1	88
		3.4.2.5 Inhibition of the *Sr*-POP-1 enzyme activity	89

4 Discussion 93

4.1 Comparison of results with published EST data from *Strongyloides* ssp. 93
4.2 Comparison with data from other parasites 98
4.3 *S. ratti* galectins 100
    4.3.1 Role of galectins in immune responses 100
    4.3.2 Galectins identified in *S. ratti* E/S products and extracts 102
4.4 The *S. ratti* POP 105
    4.4.1 Role and classification of POPs 105
    4.4.2 *Sr*-POP identified in parasitic female E/S products and extracts 106

5 Abstract 108

6 Zusammenfassung 110

7 Acknowledgements 112

8 References 113

9 Appendices 119

9.1 Protein Lists 119

9.1.1 Table 1a: List of *Strongyloides* EST cluster numbers found in E/S products from the parasitic, the infective and the free-living stages 119

9.1.2 Table 1b: Nematode RefSeq proteins found in supernatants from the parasitic, the infective and the free-living stages 124

9.1.3 Table 2a: List of *Strongyloides* EST cluster numbers found only in E/S products from infective larvae 125

9.1.4 Table 2b: Nematode RefSeq proteins only found in E/S products from the infective larvae 132

9.1.5 Table 3a: List of *Strongyloides* EST cluster numbers found only in E/S products from parasitic females 133

9.1.6 Table 3b: Nematode RefSeq proteins only found in E/S products from the parasitic females 136

9.1.7 Table 4a: List of *Strongyloides* EST cluster numbers found in E/S products from infective larvae and parasitic females 137

9.1.8 Table 4b: Nematode RefSeq proteins only found in E/S products from infective larvae and parasitic females 140

9.1.9 Table 5a: List of *Strongyloides* EST cluster numbers found only in E/S products from the free-living stages 141

9.1.10 Table 5b: Nematode RefSeq proteins only found in E/S products from the free-living stages 142

9.1.11 Table 6a: List of *Strongyloides* EST cluster numbers found in E/S products from infective larvae and free-living stages 143

9.1.12 Table 6b: Nematode RefSeq proteins found in E/S products from infective larvae and free-living stages 144

9.1.13 Table 7a: List of *Strongyloides* EST cluster numbers found in E/S products from parasitic females and free-living stages 145

9.1.14 Table 8a: List of *Strongyloides* EST cluster numbers found in extracts from the parasitic, the infective and the free-living stages 146

9.1.15 Table 8b: Nematode RefSeq proteins found in extracts from the parasitic, the infective and the free-living stages 148

9.1.16 Table 9a: List of *Strongyloides* EST cluster numbers found only in extracts from infective larvae 149

9.1.17 Table 9b: Nematode RefSeq proteins found only in extracts from infective larvae 150

9.1.18 Table 10a: List of *Strongyloides* EST cluster numbers found only in extracts from parasitic females 151

9.1.19 Table 10b: Nematode RefSeq proteins found only in extracts from parasitic females 154

9.1.20 Table 11a: List of *Strongyloides* EST cluster numbers found in extracts from infective larvae and parasitic females 155

9.1.21 Table 11b: Nematode RefSeq proteins found in extracts from infective larvae and parasitic females 156

9.1.22 Table 12a: List of *Strongyloides* EST cluster numbers found only in extracts from the free-living stages 157

9.1.23 Table 12b: Nematode RefSeq proteins found only in extracts from the free-living stages 165

9.1.24 Table 13a: List of *Strongyloides* EST cluster numbers found in extracts from infective larvae and free-living stages 166

9.1.25 Table 13b: Nematode RefSeq proteins found only in extracts from infective larvae and free-living stages 167

9.1.26 Table 14a: List of *Strongyloides* EST cluster numbers found in extracts from parasitic females and free-living stages 168

9.1.27 Table 14b: Nematode RefSeq proteins found in extracts from parasitic females and free-living stages 170

9.2 Galectin sequences 171

9.2.1 *Sr*-Gal-1 171

9.2.2 *Sr*-Gal-2 172

9.2.3 *Sr*-Gal-3 173

9.2.4 *Sr*-Gal-5 174

9.2.5 Lactose-agarose bead isolation: Identified galectin peptides 175

Abbreviations

1-D SDS PAGE	1-Dimensional sodium-dodecylsulfate polyacrylamide-electrophoresis
APS	Ammonium persulfate
BLAST	Basic local alignment search tool
CHX	Cycloheximide
Cov	Coverage
CRD	Carbohydrate recognition domain
DEPC	Diethylpyrocarbonate
DTT	Dithiothreitol
ELISA	Enzyme-linked immuno-sorbent assay
ESI	Electrospray ionisation
E/S	Excretory-secretory
EST	Expressed sequence tag
fls	Free-living stages
FTP	File transfer protocol
HEPES	4-(2-Hydroxyethyl)piperazine-1-ethanesulfonic acid
HTLV	Human T-lymphotrophic virus
i.d.	Inside diameter
IL	Interleukin
iL3	Infective third stage larvae
IPTG	Isopropyl-D-thiogalactopyranoside
L1	First stage larvae
LC-MS/MS	Liquid chromatography tandem mass spectrometry
LDS	Lithium dodecyl sulphate
Lgt	Length
LTQ	Linear triple quadrupole
mgf	Mascot generic format
MPTP	1-Methyl-4-phenyl-1,2,3,6-tetrahydropyridin
m/z	Mass to charge ratio
NCBI	National Center for Biotechnology Information
OD	Optical density
p.a.	Pro analysi
PBS	Phosphate buffered saline
PCR	Polymerase chain reaction
pf	Parasitic females
PMSF	Phenylmethanesulphonylfluoride
POP	Prolyl oligopeptidase
RT	Room temperature
STI	Swiss Tropical Institute
TBS	Tris buffered saline
TEMED	Tetramethylethylenediamine
$Th_{1/2}$	T-helper cell type 1 or 2
TIC	Total ion current
UPS	Unused protein score

Abbreviations of organisms

A. aegypti	Aedes aegypti	Yellow fever mosquito
A. ceylanicum	Ancylostoma ceylanicum	Hookworm of human and hamster
A. chlorophenolicus	Arthrobacter chlorophenolicus	Gram-positive obligate aerobe bacterium
A. elegantissima	Anthopleura elegantissima	Anemoe
A. irradicans	Argopecten irradicans	Species of saltwater clam
A. lumbricoides	Ascaris lumbricoides	Roundworm, nematode parasite of humans and animals
A. mellifera	Apis mellifera	European honey bee
A. pisum	Acyrthosiphon pisum	Aphid
A. proteobacterium BAL199	Alpha proteobacterium BAL199	Proteobacterium
A. suum	Ascaris suum	Parasitic nematode of pigs
A. thaliana	Arabidopsis thaliana	Small flowering plant
A. trivirgatus	Aotus trivirgatus	Species of owl monkey
A. vitae	Acanthocheilonema vitae	Rodent filarial nematode
B. elongata	Barentsia elongata	Entoproct, phylum of small aquatic animals
B. malayi	Brugia malayi	Filarial parasite of humans
B. mori	Bombyx mori	Silk moth
B. pahangi	Brugia pahangi	Filarial parasite of cats
B. xylophilus	Bursaphelencus xylophilus	Pine wood nematode
C. beijerinckii	Clostridium beijerinckii	Gram-positive bacterium
C. botulinum	Clostridium botulinum	Gram-positive bacterium
C. brenneri	Caenorhabditis brenneri	Non-parasitic nematode
C. briggsae	Caenorhabditis briggsae	Non-parasitic nematode
C. elegans	Caenorhabditis elegans	Non-parasitic nematode
C. familiaris	Canis familiaris	Domestic dog
C. intestinalis	Ciona intestinalis	Sea squirt
C. pipiens quinquefasciatus	Culex pipiens quinquefasciatus	Mosquito, vector of human pathogens
C. reinhardtii	Chlamydomonas reinhardtii	Motile single celled green alga
C. remanei	Caenorhabditis remanei	Non-parasitic nematode
D. ananassae	Drosophila ananassae	Fruit fly
D. citri	Diaphorina citri	Psyllid feeding on citrus plants
D. destructor	Ditylenchus destructor	Plant pathogenic nematode
D. discoideum AX4	Dictyostelium discoideum AX4	Soil-dwelling social amoeba strain AX$
D. grimshawi	Drosophila grimshawi	Fruit fly
D. immitis	Dirofilaria immitis	Filarial parasite of dogs
D. melanogaster	Drosophila melanogaster	Fruit fly
D. rerio	Danio rerio	Zebrafish
D. simulans	Drosophila simulans	Fruit fly
D. viviparus	Dictyocaulus viviparus	Parasitic lungworm of cattle
D. willistoni	Drosophila willistoni	Fruit fly
D. yakuba	Drosophila yakuba	Fruit fly
E. coli	Escherichia coli	Gram-negative bacterium
E. dispar	Entamoeba dispar	Protozoal parasite

E. ventriosum	Eubacterium ventriosum	Anaerobic, gram positive bacterium
F. foliacea	Flustra foliacea	Sea mats
G. gallus	Gallus gallus	Domesticated fowl, chicken
G. rostochiensis	Globodera rostochiensis	Plant parasitic nematode
H. contortus	Haemonchus contortus	Parasitic nematode of ruminants
H. glycines	Heterodera glycines	Plant parasitic nematode
H. sapiens	Homo sapiens	Human
H. virescens	Heliothis virescens	Moth species
K. sp. RS1982	Koerneria species strain	Nematode
L. loa	Loa loa	Parasitic filarial nematode of humans
L. major	Leishmania major	Protozoal intracellular parasite of humans
L. obliqua	Lonomia obliqua	Moth species
L. vannamei	Litopenaeus vannamei	A variety of prawn
L. vestfoldensis	Loktanella vestfoldensis	Proteobacterium
M. incognita	Meloidogyne incognita	Plant parasitic nematode
M. mulatta	Macaca mulatta	Rhesus monkey
M. musculus	Mus musculus	Mouse
N. americanus	Necator americanus	Parasitic nematode of mammals
N. vectensis	Nematostella vectensis	Species of sea anemone
N. vitripennis	Nasonia vitripennis	Small parasitoid wasp
O. ostertagi	Ostertagia ostertagi	Nematode parasite of cattle
O. sativa	Oryza sativa	Rice
O. tauri	Ostreococcus tauri	Unicellular green alga
O. volvulus	Onchocerca volvulus	Filarial parasite of humans
P. caudatus	Priapulus caudatus	Marine worm
P. marneffei	Penicillium marnefei	Human pathogenic fungus
P. maupasi	Pristionchus maupasi	Non-parasitic nematode
P. sp. 3 CZ3975	Novel Pristionchus species	Non-parasitic nematode
P. sp.6 RS5101	Novel Pristionchus species	Non-parasitic nematode
P. tetraurelia	Paramecium tetraurelia	Unicellular ciliate protozoa
P. trichosuri	Parastrongyloides trichosuri	Nematode parasite of mammals
R. torques	Ruminococcus torques	Gram-positive bacterium of ruminants
P. troglodytes	Pan troglodytes	Chimpanzee
R. etli	Rhizobium etli	Gram-negative bacterium
R. norvegicus	Rattus norvegicus	Norway rat
S. salar	Salmo salar	Atlantic salmon
S. bicolor	Sorghum bicolor	Poaceae, plant species
S. cephaloptera	Spadella cephaloptera	Predatory marine worm
S. coelicolor	Streptomyces coelicolor	Gram-positive actinobacterium
S. feltiae	Steinernema feltiae	Parasitic nematode
S. pneumoniae	Streptococcus pneumoniae	Gram-positive, human pathogenic bacterium
S. scrofa	Sus scrofa	Boar
S. fuelleborni	Strongyloides fuelleborni	Nematode parasite of non-human primates and humans
S. papillosus	Strongyloides papillosus	Nematode parasite of ruminants and rabbits
S. ratti	Strongyloides ratti	Nematode parasite of rats
S. stercoralis	Strongyloides stercoralis	Nematode parasite of humans
S. trachea	Syngamus trachea	Nematode parasite of fowl and wild birds

Abbreviations

S. sviceus	*Streptomyces sviceus*	Gram-positive bacterium
T. brucei	*Trypanosoma brucei*	Protozoan parasite of humans
T. circumcincta	*Teladorsagia circumcincta*	Nematode parasite of sheep and goats
T. canis	*Toxocara canis*	Nematode parasite of dogs
T. castaneum	*Tribolium castaneum*	Red flour beetle
T. cruzi	*Trypanosoma cruzi*	Protozoan parasite of humans
T. denticola	*Treponema denticola*	Gram negative, highly proteolytic bacterium
T. gondii	*Toxoplasma gondii*	Parasitic protozoa
T. nigroviridis	*Tetraodon nigroviridis*	Green spotted pufferfish
T.spiralis	*Trichinella spiralis*	Nematode parasite of mammals
T. pseudospiralis	*Trichinella pseudospiralis*	Nematode parasite of mammals
T. trichiura	*Trichuris trichiura*	Whipworm, nematode parasite of humans
T. vaginalis	*Trichomonas vaginalis*	Anaerobic, parasitic flagellated protozoan
W. bancrofti	*Wuchereria bancrofti*	Filarial parasite of humans
X. tropicalis	*Xenopus tropicalis*	Frog species
Y. lipolytica	*Yarrowia lipolytica*	Yeast

List of figures

1.1-*1*	Dendrogram for the phylum nematoda	1
1.1-*2*	The life cycle of *S. stercoralis*	2
1.7-*1*	Secretion mechanisms of cytosolic proteins	9
1.7-*2*	Excretory/secretory system of nematodes	10
2.3.1.2-*1*	Charcoal culture and Baermann apparatus	20
2.3.3.10-*1*	Setup of the LTQ MS	32
2.3.3.10-*2*	The formation of peptide ions by ESI	33
2.3.3.10-*3*	Fragmentation mechanism of a doubly charged peptide ion	33
2.3.3.10-*4*	Xcalibur® Qual Browser window	34
2.3.3.10-*5*	The ProteinPilot™ user interface	35
2.3.3.10-*6*	Protein composition of a single gel band	36
3.1-*1*	Time course of a *S. ratti* infection	42
3.1-*2*	Development of different free-living stages during a *S. ratti* infection	42
3.1-*3*	Rapid development of the life cycle	43
3.2.1-*1*	Amount of E/S products depends on number of infected larvae	44
3.2.2-*1*	Inhibition of the metalloprotease	45
3.2.3-*1*	Gelatin overlay for size determination of the protease	46
3.2.4-*1*	Coomassie and silver stain of iL3 E/S products	46
3.2.4-*2*	iL3 E/S products incubated over different time periods	47
3.2.5-*1*	iL3 E/S products incubated at different temperatures	48
3.2.6-*1*	Different treatment methods for the incubation of iL3	48
3.2.7-*1*	1-D SDS PAGE of iL3, pf and fls E/S products and extracts	49
3.2.8-*1*	iL3, pf and fls E/S products show distinct band patterns	50
3.2.9-*1*	ELISA shows the antibody recognition to iL3 E/S products	51
3.3-*1*	General workflow for mass spectrometric analysis	52
3.3.1-*1*	Mass spectrometry results of whole iL3 analysis	53

List of figures   XIII

3.3.1-2	Venn diagram of protein distribution found in E/S products	54
3.3.1-3	Venn diagram of protein distribution found in extracts	54
3.3.1-4	Schematic representation of proteins within single stages	55
3.3.1-5	Schematic representation of proteins within E/S- and extract proteins	56
3.3.4-1	PCR analysis shows differential protein expression	66
3.4.1-1	PCR analysis of all galectins found in the *S. ratti* EST database	68
3.4.1.1-1	Galectins carrying the SL-1 sequence	69
3.4.1.2-1	Scheme of the domain structure of the *S. ratti* galectins	70
3.4.1.2-2	Multiple sequence alignment of *Sr*-Gal-1, -2, -3 and -5	71
3.4.1.2-3	Tertiary structure prediction of the two CRD regions from *Sr*-Gal-3	72
3.4.1.3-1	Phylogenetic tree of nematode galectins	73
3.4.1.4-1	Chair form of lactose	74
3.4.1.4-2	Sequence coverages of *Sr*-Gal-1, -2 and -3 obtained in affinity separation	75
3.4.1.5-1	Purification of recombinant *Sr*-Gal-3	76
3.4.1.6-1	Antibody recognition of recombinant *Sr*-Gal-3	77
3.4.1.7-1	Principle of carbohydrate microarrays	78
3.4.1.7-2	Carbohydrate microarray of a *S. ratti* iL3 extract	79
3.4.1.7-3	Carbohydrate structures and molecular formulas bound to array	79
3.4.2.1-1	Overlapping N- and C-terminal ends of POP cluster sequences	80
3.4.2.1-2	RACE PCR for the completion of the *Sr*-POP-1 sequence	81
3.4.2.2-1	Sequence coverage of *Sr*-POP-1 peptides	82
3.4.2.3-1	Full length nucleotide sequence of *Sr*-POP-1	84
3.4.2.3-2	Domain structure of *Sr*-POP-1	85
3.4.2.3-3	Sequence logo of the serine active site from the POP family	85
3.4.2.3-4	Ribbon diagram of *Sr*-POP-1	86
3.4.2.4-1	Sequence alignment of C-terminal *Sr*-POP-1 ends	88
3.4.2.4-2	Phylogenetic tree of selected POPs	88

3.4.2.5-*1*	Structures of POP inhibitors	89
3.4.2.5-*2*	Effect of different POP inhibitors during *in vitro* culture	91
4.3.1-*1*	Galectin structures	99
4.4.1-*1*	Classification of peptidases/proteinases	104

List of tables

2.2.1-*1*	Devices used at the BNI, Hamburg	12
2.2.1-*2*	Devices used at the Proteomics Center, Boston	13
2.2.2-*1*	Kits used in the laboratory	13
2.2.6-*1*	Oligonucleotide sequences	17
2.3.3.2-*1*	Composition of SDS gels	29
2.3.3-*1*	Mass shift of fragment ions from homologous peptides	36
2.3.4.1-*1*	Composition of the search database	37
3.1-*1*	Scheme for the maintenance of the *S. ratti* life cycle	41
3.3.2-*1*	25 highest scoring proteins found in all stages	59
3.3.3.1-*1*	25 highest scoring proteins of iL3 E/S products	61
3.3.3.2-*1*	25 highest scoring proteins of pf E/S products	63
3.3.3.3-*1*	25 highest scoring proteins of fls E/S products	64
3.4.1-*1*	*S. ratti* and *S. stercoralis* EST clusters homologous to galectins	67
3.4.1-*2*	Pairwise comparison of *S. ratti* and *S. stercoralis* galectin sequences	68
3.4.1.4-*1*	Protein sequences analysed in the bound fraction eluate 3	75
3.4.2.2-*1*	Identified peptides for *Sr*-POP-1	82
3.4.2.4-*1*	Homologies of the *Sr*-POP-1 with related proteins from selected species	87
4.1-*1*	Proteins having a significantly higher expression in parasitic females subject to low or high immune responses	97
4.3-*1*	Seven *S. ratti* galectin sequences and the corresponding cluster numbers	101

1 Introduction

In the presented work the helminth parasite *Strongyloides ratti* was used as a model nematode to study the composition of secreted and extract proteins to gain deeper insights into molecules that are important for the establishment and maintenance of parasitism.

1.1 The life cycle of *Strongyloides* spp.

Parasitic nematodes are widespread and important pathogens of humans, animals and plants. It is estimated that more than a quarter of the world's human population is infected with nematodes. The majority of the exposed population lives in the developing world (Awasthi, 2003; World Health Organization, 2003). Referring to their final habitat within their natural hosts, parasitic helminths can be generally divided in two major subgroups – irrespective of the way of infection: tissue dwelling and gastrointestinal helminths. *S. ratti* can be allocated to the latter group of helminths and belongs to the phylum nematoda. Within the phylogeny of nematodes *S. ratti* can be found in the order Rhabditida, family *Strongyloididae* (Figure 1.1-*1*).

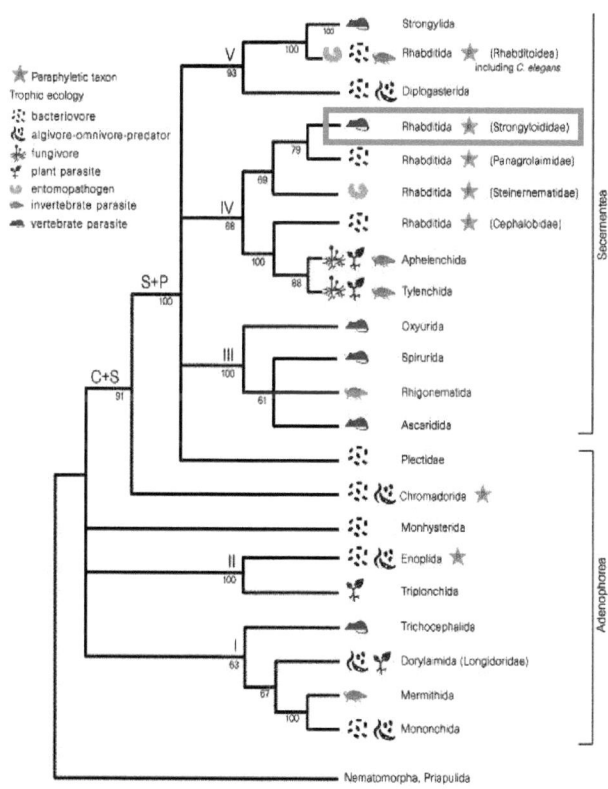

Figure 1.1-*1* Dendrogram for the phylum nematoda (from Blaxter, 1998)

The family of *Strongyloididae* covers roughly 60 different species and subspecies of which 22 can be found on the taxonomy browser of the National Center for Biotechnology Information (NCBI) homepage (www.ncbi.nlm.nih.gov/sites/entrez?db=taxonomy). They all infect different vertebrate hosts and share parasitic and free-living adult generations which makes their developmental life cycle unique among nematode parasites of vertebrates (Figure 1.1-2). Here the *Strongyloides stercoralis* life cycle in humans is presented which is comparable to the *S. ratti* life cycle in rats.

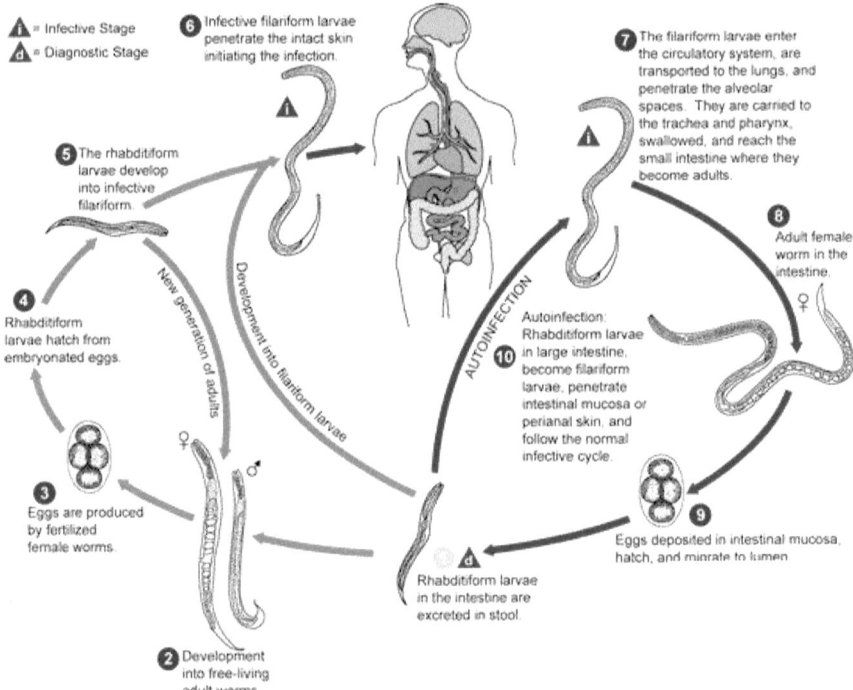

Figure 1.1-2 The life cycle of *S. stercoralis* (Centers for Disease Control and Prevention, Division of Parasitic Diseases, http://www.dpd.cdc.gov/dpdx/)

Infective third stage larvae (iL3) can persist in the environment until they encounter a suitable host. They then enter the host body by skin penetration. The larvae proceed by means of tissue migration until reaching a blood vessel. Via the circulation they migrate to the lungs where they break out of the pulmonary capillaries and enter the alveoli (Schad, 1989). After travelling up the respiratory tree and entering the pharynx they are swallowed and finally reach the small intestine. During this process the iL3 develop into fourth stage larvae which develop further into the adult parasitic generation. This intestinal phase of infection is constituted of females only that reproduce by parthenogenesis (Viney, 1994), a form of asexual reproduction in which females produce eggs that develop without fertilisation. The parasites lie embedded in the mucosal epithelium of the small intestine where they deposit their eggs. These eggs can undergo three pathways of development. Firstly they can develop directly into autoinfective third stage larvae which ensures that *Strongyloides* can persist in its host for long time periods or even a lifetime (Hauber, 2005). Secondly the eggs are passed out of the host in faeces. However some of the first stage larvae (L1) can already hatch before the eggs reach the external environment. The eggs are either male or female (Harvey, 2001). Male eggs hatch and can only develop into free-living adult males by moulting through four larval stages. The female eggs, in contrast, have a developmental choice. They can either develop through four larval stages into free-living adult females - termed heterogonic or indirect development - or alternatively they can develop through two larval stages directly into iL3s - termed homogonic or direct development. Following the mating by sexual reproduction of the free-living adults the females lay eggs. To complete the life cycle the eggs hatch and develop through two larval stages into iL3s.

1.2 Pathology of *Strongyloides* infections

S. stercoralis and *Strongyloides fuelleborni* are two species infecting humans (Ashford, 1989). It is estimated that the worldwide prevalence of infected people is 50–200 million (Compton, 1987; Albonico, 1999), making it the fourth most important intestinal nematode infection, after hookworm, *Ascaris lumbricoides* and *Trichuris trichiura* (Stephenson, 2000).

Strongyloidiasis is considered a systemic infection although the parasite is an intestinal nematode. In many cases, mostly in hosts with normal immune status the infection proceeds asymptomatically, thus the parasite cycle can persist undetected for decades. Several organs and tissues e.g. the hepatobiliary system, the pancreas and skin can show abnormalities during a *Strongyloides* infection. However gastrointestinal symptoms are the most common and the respiratory tract is the system most frequently affected outside the gastrointestinal tract. Therefore, most common symptoms are progressive weight loss, diarrhoea, abdominal pain and vomiting. Also noticeable are dermatologic signs as skin rash and *larva currens* and alterations in blood count due to eosinophilia which is present in more than 70% of the cases.

Hyperinfection and dissemination are two terms commonly used to denote the severe form of strongyloidiasis. Triggers include immunosuppressive therapy, human T-lymphotrophic virus-1 (HTLV) infection, hematologic malignant disease and transplantation. The term „hyperinfection" describes an increase in the rate of autoinfection, when the worms are detectable in extraintestinal regions, especially the lungs, due to a rapid and overwhelming penetration of iL3 through the intestinal wall. The term "disseminated" is usually restricted to infections in which worms can be found in ectopic sites, e.g. the brain. In both stages a complete disruption of the mucosal patterns, ulcerations, and paralytic ileus has been observed. Bacterial and fungal infections often occur in cases of hyperinfection because of the leakage of gut flora from a bowel damaged by moving larvae. As a result the infection leads to severe pathological symptoms that, in addition to the above mentioned symptoms include dyspnoea, haemoptysis, cough, respiratory distress and fever (Genta, 1989). The infection then can take a fatal outcome in the absence of therapy or if diagnosed and treated too late (Viney, 2004; Lim, 2004).

1.3 Treatment of strongyloidiasis

In the developed countries helminth infections, including strongyloidiasis in humans, can widely be controlled through primary health care programs and effective public sanitation whereas in developing nations helminth diseases are still widespread and often drug treatment does not protect against rapid re-infection (Anthony, 2007).

In general the options for the treatment of helminth infections offer a range of different active substances e.g. benzimidazoles, macrocyclic lactones, tetrahydrompyrimidines and emodepsides.

The drug of choice for strongyloidiasis is the macrocyclic lactone ivermectin (www.dpd.cdc.gov. 2008) which is derived from the bacterium *Streptomyces avermitilis* (Li, 2008). It binds to and activates glutamate-gated chloride channels which can be predominantly found in neurons and myocytes of non-vertebrates. This leads to an influx of calcium ions causing hyperpolarisation of the cell membrane and ultimately death.

Albendazole, belonging to the chemical class of benzimidazoles, is the recommended alternative (www.dpd.cdc.gov. 2008) to ivermectin for the treatment for an infection with *Strongyloides*. Its proposed mechanism of action is the inhibition of tubulin polymerisation in intestinal parasites which leads to metabolic interception including the loss of energy metabolism. These pathophysiological alterations lead to parasite death. Both drugs can be administered orally in non severe strongyloidiasis whereas in hyperinfection syndrome a combination therapy has been proposed (Lim, 2004).

Other therapeutic agents like diethylcarbamazine, the mode of action of which presumably lies in its gabaergic and cholinergic effect on the parasite's central nervous system (Terada, 1985), have been described (Harder, 2002). The symmetrical diamidine derivative tribendimidine, a new anthelmintic agent which has been approved for human use by Chinese authorities in 2004 has recently been tested against *S. stercoralis* in an open label randomized trial compared to albendazole (Steinmann, 2008). The mentioned study tested the effect of a single dose and showed a slight reduction in the worm load. However a multiple-dose study with tribendimidine is still outstanding to show if this drug has satisfactory effects against *S. stercoralis*. Other than tribendimidine most of the before mentioned active substances were introduced into the market many years ago and were widely used in humans and animals. The resulting reduced efficacy of common anthelmintic drugs in veterinary medicine shows the need for the development of new therapeutic agents for the treatment of parasitic nematode infections (Kaplan, 2004).

1.4 Diagnosis of strongyloidiasis

The diagnosis of strongyloidiasis can only be made in a laboratory because the only pathognomic clinical sign of a *Strongyloides* infection is the *larva currens* which does not necessarily occur. A PCR method with specificity for *S. stercoralis* has been developed at the BNI, however, the validation of the sensitivity is still in process. It is therefore recommended to examine repeated stool samples over a number of consecutive days, which is overall the best method (World Gastroenterology Organisation: Practice Guideline - Management of strongyloidiasis, 2004). The following tests methods for the examination of stool samples can be applied:

- Baermann technique:

    See section 2.3.1.2

- Harada-Mori filter paper technique:

    A faecal sample is smeared on a filter paper strip leaving 5 cm clear at one end. The strip is inserted into a test tube filled with a few mL of distilled water with the unsmeared portion reaching into the water. The test tube is sealed and incubated four days at 28°C. Infective larvae will be visible in the water under the microscope.

- Koga agar plate method:

  The stool is placed on agar plates. After two days at room temperature larvae crawl over the surface and carry bacteria with them, creating visible tracks.

- Direct staining of faeces:

  Faecal smears can be stained with saline Lugol iodine solution or auramine O. However, single stool examination detects larvae in only 30% of cases of infection.

Other diagnostic tools include the detection of anti-*Strongyloides* antibody using enzyme linked immuno-sorbent assay (ELISA) technique in case *Strongyloides* antigen is available. However, the ELISA test cannot distinguish antibodies produced of past or current infections. If the described methods fail to detect *Strongyloides* an endoscopy can reveal mucosal erythema and edema in the duodenum and a specimen of duodenal fluid can contain both, eggs and larvae.

1.5   Immune response to *Strongyloides*

Generally infectious agents can be divided into micropathogens, including viruses, bacteria and protozoa and macropathogens, including helminths. Unlike most micropathogenic infections, which are acute and short-lived, macropathogenic infections are long-lasting, chronic infections (Maizels, 1993). The host immune response is the result of a prolonged dynamic co-evolution between the host and the parasite. For *S. ratti* in its rat host the immune response results in the reduction of the size of the parasitic female stages, a reduction in their *per capita* fecundity, the adoption of a more posterior position in the host gut and, ultimately, the death of these stages (Wilkes, 2007). In addition, the host immune response also affects the developmental route of the free-living stages of the *S. ratti* life cycle. As an infection progresses the number of larvae that develop into free-living males increases. At the same time the proportion of female larvae that develop into free-living females and directly developing iL3s also increases (Harvey, 2000). Thus the host immune response plays an important role in controlling *Strongyloides* infections.

Mostly helminth parasite infections result in a shift from a type 1 T-helper cell (Th_1) to a type 2 T-helper cell (Th_2) immune response which has also been observed in *S. ratti* infections (Bleay, 2007). Th_2-type responses are characterised by increased levels of interleukin-4 (IL-4) and other Th_2-type cytokines (including IL-5, IL-9, IL-13 and IL-21), by activation and expansion of CD4+ Th_2 cells and plasma cells secreting IgE, by eosinophils, mast cells and basophils, all of which can produce several types of Th_2-type cytokines. This characteristic shift to Th_2-type responses has also been observed in *Strongyloides* infections (Wilkes, 2007; Porto, 2001). In *S. ratti* infections of rats circulating anti-*S. ratti* IgG_1 and IgG_{2a} response has been detected and at the same time it was found that there is a greater IgG response to parasitic females than to iL3 (Wilkes, 2007). In intestinal tissue a specific anti-*S. ratti* IgA response which increases during the infection has been observed. IgG responses have also been observed in humans infected with *S. stercoralis* (Viney, 2004). Also the increase of the eosinophil count in peripheral blood is a common feature in human *S. stercoralis* infection together with increased levels of IgG, IgE and eotaxin. Eotaxin is a Th_2 chemokine with preferential chemotactic activity for eosinophils and is also involved in eosinophil-associated inflammatory responses, such as allergic diseases, haematological diseases and in inflammatory bowel disease. It is expressed constitutively in the gastrointestinal tract, where it is a fundamental regulator of the physiological trafficking of eosinophils during healthy states, and where it is believed it may be responsible for host defence against parasites (Mir, 2006).

1.6 *Strongyloides ratti* as a laboratory model

It has been shown in the laboratory that the human parasite *S. stercoralis* can be maintained in dogs (Lok, 2007) and gerbils (Nolan, 1999). The use of dogs, however, is prohibitively expensive and may be a source of ethical concern. It has to be considered that neither dogs nor gerbils are the natural hosts for *S. stercoralis*. Though, when using laboratory models of nematode infection the natural infection conditions should accurately be imitated in order to use the resulting information to understand human infections.

The rat-invading parasite *S. ratti* comprises various features which make it an ideal organism to work with in the laboratory. As mentioned above the life cycle consists of a parasitic and a non-parasitic phase thus, in contrast to most other nematodes, there is no insect vector needed for the perpetuation of the cycle. The absence of an insect vector is important from two points of view. Firstly it allows comparing parasitic and non-parasitic stages from the same species on the molecular level in order to evaluate genes and gene products that might be important in the process of infection and in parasite survival within its host. Also the development of the respective stages and secondly the handling is less time-consuming.

1.7 Protein secretion of nematodes

- Cellular secretion mechanisms

Excretory/secretory proteins of parasites were subject of intensive studies during the last decades (Lightowlers, 1988) and represent an integral part of the presented work. Thus it is important to explain in detail the various known mechanisms of protein transport out of the cell and out of the worm.

In the classical secretion pathway of soluble proteins or the transport of membrane proteins to the cell surface it is required that the proteins are processed into and through the endoplasmatic reticulum and the Golgi apparatus. Usually signal peptides or transmembrane domains target proteins for translocation into the lumen or insertion into the membrane of the endoplasmatic reticulum.

Non-classical secretion pathways of proteins lacking a signal sequence include plasma membrane translocation, ectocytosis, autophagy and intracellular vesicular transport (Fig. 1.7-*1*).

In translocation (Figure 1.7-*1* A) the secretion of cytosolic proteins is mediated by transporters, notably the superfamily of ATP-binding cassette (ABC) transporters. ABC proteins which are integrated into the membrane are capable of translocating a wide range of molecules including proteins across membranes (Schatz, 1996).

In intracellular vesicular transport (Figure 1.7-*1* B) cytosolic proteins are first incorporated into intracellular vesicles, then, released into the extracellular space as free components upon fusion of the vesicles with the plasma membrane. The origin of these vesicles is not clear but they possibly derive from the trans-Golgi-network. However, it remains unclear how a cytosolic protein lacking a signal sequence that is in general being required for trans-Golgi-network transport could enter into vesicular trafficking at this late stage (Traub, 1997).

In autophagy (Figure 1.7-*1* C) inclusion vesicles are formed by invagination of the membrane of a large intracellular endosome. These protein containing inclusion vesicles are then externalised by fusion of the membrane of the multivesicular compartment with the plasma membrane (Klionsky, 1998).

In ectocytosis (Figure 1.7-1 D) cytosolic proteins concentrate in the cytoplasm underlying plasma membrane domains. The membrane then forms protrusions („blebs') including the previously formed protein aggregates. The blebs finally detach from the plasma membrane and are released as extracellular vesicles from which soluble proteins are released (Beaudoin, 1991).

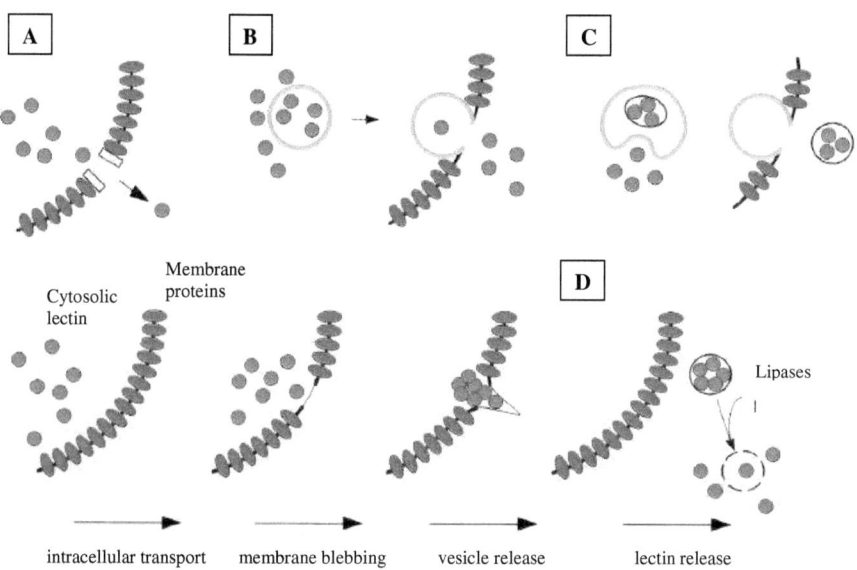

Figure 1.7-1 Secretion mechanisms of cytosolic proteins (Hughes, 1999). (A) Plasma membrane translocation, (B) intracellular vesicular transport, (C) autophagy, (D) ectocytosis.

- Nematode excretory system

Parasitic nematodes possess specialised excretory and secretory organs. The E/S system of nematodes is a unique structure exhibiting variable morphology among different nematode species. It was first named through morphological descriptions and its function is osmoregulatory, ion regulatory, secretory and excretory. The secretions of parasitic nematodes are generally referred to as E/S products. The structure of the E/S system of the free-living nematode *C. elegans* has been subject of intensive studies (Thompson, 2002) and is exemplary presented in this work.

The body of *C. elegans* and many other nematodes, including *S. ratti* and *S. stercoralis*, is permeated with excretory canals (Figure 1.7-2).

These canals are part of a single H-shaped, extremely large canal cell lying behind the isthmus of the pharynx. The canal cell together with two other cell types of the E/S system, the duct cell and the pore cell, is fused to two excretory gland cells. The excretory canal cell functions in part as a kidney, excreting saline fluid via the duct and pore in order to maintain the animal's salt balance (osmoregulation) and probably to remove metabolites (Buechner, 1999). The excretory gland cell is connected to the same duct and pore and secretes materials from large membrane-bound vesicles. The nature of this secretion is unknown, but has been postulated to play a role in moulting and for parasitic nematodes to be the source of molecules which are antigenic in infected hosts or possibly immunomodulatory involved in parasite-host-interaction (Lightowlers, 1988). Therefore the identification of proteins and analysis of the composition of the components secreted by parasitic nematodes can contribute to the understanding of the mechanisms responsible for the infection and perpetuation of the parasite within its respective host.

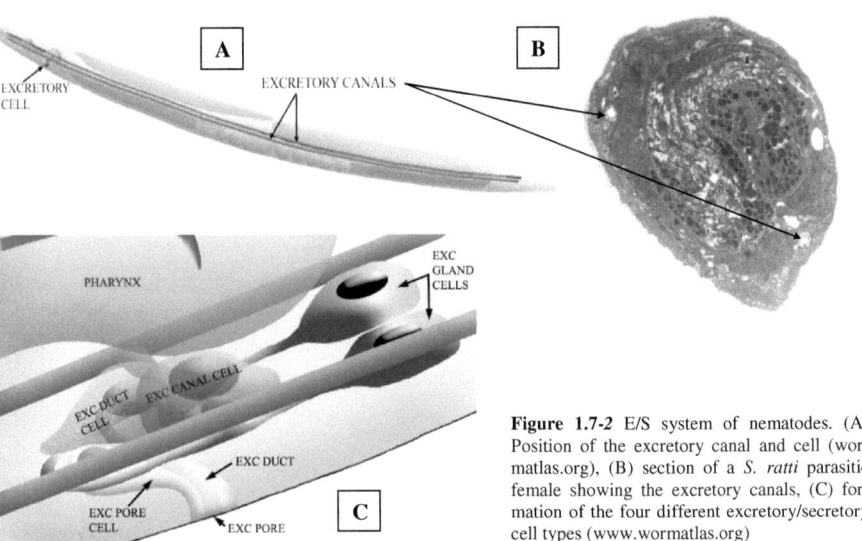

Figure 1.7-2 E/S system of nematodes. (A) Position of the excretory canal and cell (wormatlas.org), (B) section of a *S. ratti* parasitic female showing the excretory canals, (C) formation of the four different excretory/secretory cell types (www.wormatlas.org)

1.8 Objectives

The aim of the thesis was to identify and describe E/S proteins from the parasitic nematode *S. ratti* which are abundantly or differentially produced by various developmental stages occurring in the life cycle: the infective larvae, parasitic females and the free-living stages. To address this task the following approach was applied:

- Set up of the *S. ratti* life cycle in *Rattus norvegicus*
- Development of procedures to obtain high numbers of different developmental stages
- Optimising of procedures to obtain excretory/secretory products and extract proteins of the respective developmental stages
- Protein analysis
    - 1-dimensional sodium-dodecylsulfate polyacrylamide-electrophoresis (1-D SDS PAGE)
    - Sequencing by liquid chromatography tandem mass spectrometry (LC-MS/MS)
    - Bioinformatic evaluation
- Functional analysis of the proteins and serological assays
    - Proteolytic activity
    - ELISA
    - Western blot
- Identification of gene sequences encoding the identified proteins and recombinant expression
    - Preparation of RNA and cDNA
    - Completion of EST sequences
    - Recombinant expression of one of the proteins

2 Animals, Materials and Methods

2.1 Animals

Rats:

For the establishment of the parasite cycle and for the preparation of different *S. ratti* life cycle stages exclusively male Wistar rats (*R. norvegicus*) aged four weeks or older were used.

Parasites:

The *S. ratti* life cycle has been held at the Bernhard Nocht Institute for Tropical Medicine (BNI, Hamburg) since January 2006. The iL3 for the initial infection were kindly supplied by Prof. Dr. Gerd Pluschke from the Swiss Tropical Institute (STI), Department of Medical Parasitology and Infection Biology, Basel, Switzerland.

2.2 Materials

2.2.1 Devices

The majority of the laboratory work for the presented thesis was conducted at the BNI. All LC-MS/MS analyses were performed at the Proteomics Center at Childrens' Hospital (CHB), Boston, MA/USA. Table 2.2.1-*1* lists the devices used at the BNI and table 2.2.1-*2* lists the devices used at the Proteomics Center. Unless otherwise stated all organic solvents, acids, bases and solid compounds were used in p.a. quality and were supplied by Merk KGaA, Darmstadt or Carl Roth GmbH, Karlsruhe for the work at the BNI. Reagents for the work at CHB were purchased from Sigma-Aldrich and HPLC-grade solvents from Burdick and Jackson.

Table 2.2.1-*1* Devices used at the BNI, Hamburg

Type	Manufacturer / Supplier
Amicon Ultra-4/-15 Ultracel-10k	Millipore
Analytical balance PT1200	Sartorius (Göttingen)
Biophotometer	Eppendorf (Hamburg)
Precellys steel beads	Peqlab (Erlangen)
Vortex MS 1 Minishaker	IKA (Staufen)
Microcon Ultracel YM-10	Millipore (Billerica)
Microcentrifuge 5415 C	Eppendorf

Animals, Materials and Methods 13

Table 2.2.1-*1* continued

Type	Manufacturer / Supplier
Power supply unit Power Pac 300	Bio Rad (Munich)
Electrophoresis chamber Perfect Blue Mini	Peqlab
Cooling centrifuge Rotanda/RP	Hettich (Tuttlingen)
Microscope Axiovert 25	Zeiss (Jena)
Stereozoom microscope Wild M8	Leica (Wetzlar)
Heating block thermomixer comfort	Eppendorf
pH meter CG 480	Schott (Mainz)
Shaking incubator Innova 4400	New Brunswick Scientific (Nürtingen)
Sterile working bench Microflow	Nunc (Wiesbaden)
Thermocycler Primus 25	Peqlab
Omnifix-F 1 mL syringes	B. Braun (Melsungen)
Sterican hypodermic needles, 0.40 x 25 mm	B. Braun
Activated charcoal, 2.5 mm granules	Merck (Darmstadt)
Powershot A95	Canon (Krefeld)

Table 2.2.1-*2* Devices used at the Proteomics Center, Boston

Type	Manufacturer / Supplier
NuPage®Novex®tris acetate mini gels	Invitrogen (Carlsbad)
NuPage®LDS sample buffer 4x	Invitrogen
SimplyBlue™safe stain	Invitrogen
XCell Sure Lock™mini cell	Invitrogen
Linear ion trap mass spectrometer - LTQ	Thermo Scientific (Waltham)
SpeedVac Concentrator 5301	Eppendorf

2.2.2 Kits

Table 2.2.2-*1* Kits used in the laboratory

Type	Manufacturer / Supplier
Bio-Rad Protein Assay	Bio-Rad
GeneRacer Kit	Invitrogen
QIAprep Spin Miniprep Kit	Qiagen (Hilden)
Illustra™ GFX™ PCR DNA and Gel Band Purification Kit	GE Healthcare (Buckinghamshire)
ProtoScript® First Strand cDNA Synthesis Kit	NEB
SuperScript™ III Reverse Transcriptase	Invitrogen

2.2.3 Solutions

Acrylamide solution 30 % (w/v)	Acrylamide/Bisacrylamide	(29:1)
Loading buffer for agarose gels	Na_2-EDTA	100 mM
	Glycerine	30 % (v/v)
	Bromphenol blue	0.05 % (w/v)
Coomassie staining solution (Hamburg)	Ethanol	50 % (v/v)
	Coomassie brilliant blue	0.05 % (w/v)
	Acetic acid	10 % (v/v)
Coomassie destaining solution (Hamburg)	Ethanol	45 % (v/v)
	Glacial acetic acid	10 % (v/v)
Gel drying solution	Ethanol	20 % (v/v)
	Glycerine	10 % (v/v)
His tag purification buffer B	Urea, pH 8.1	8 M
	Tris-HCl	10 mM
	$NaH_2PO_4 \cdot H_2O$	100 mM
His tag purification buffer C	Urea, pH 6.3	8 M
	Tris-HCl	10 mM
	NaH_2PO_4	100 mM
His tag purification buffer D	Urea, pH 5.9	8 M
	Tris-HCl	10 mM
	NaH_2PO_4	100 mM
His tag purification buffer D	Urea, pH 4.5	8 M
	Tris-HCl	10 mM
	NaH_2PO_4	100 mM
IPTG stock solution	IPTG	1 M
PBS 10x	Na_2HPO_4	80.6 mM
	KH_2PO_4	14.7 mM
	NaCl	1.37 M
	KCl	26.8 mM
	pH 7,0 at 25°C	
SDS PAGE running buffer (10x)	Tris-HCl	250 mM
	Glycine	1.9 M
	SDS	1 % (v/v)

SDS PAGE sample buffer (6x)	Tris-HCl Glycerol SDS Dithiothreitol Bromphenol blue pH 6,8 by addition of HCl	0.125 M 25 % (v/v) 2 % (v/v) 50 mM 0.02 % (w/v)
SDS PAGE stacking gel buffer (4x Tris/SDS pH 6,8)	Tris-HCl SDS pH 6,8 by addition of HCl	500 mM 0.4 % (v/v)
SDS PAGE separation gel buffer (4x Tris/SDS pH 8,8)	Trisbase SDS pH 8,8 adjust with HCl	1.5 M 0.4 % (v/v)
Silver stain: Fixing solution I	Ethanol Acetic acid	30 % (v/v) 10 % (v/v)
Silver stain: Fixing solution II	Ethanol Sodium acetate Glutaric aldehyde $Na_2S_2O_3$	30 % (v/v) 0.5 M 25 % (v/v) 0.2 % (w/v)
Silver stain: Staining solution IV	Silver nitrate Formic aldehyde	0.1 % (w/v) 0.01 % (v/v)
Silver stain: Developer	Sodium carbonate Formic aldehyde pH 11.3-11.8 at 25°C	2.5 % (w/v) 0.01 % (v/v)
Renaturation buffer	Tris-HCl NaCl pH 7.5 at 25°C	50 mM 100 mM
TAE (50x)	Tris Acetic acid EDTA pH 8.3 at 25°C	2 M 1 M 50 mM
TFB I	Calcium chloride Glycerol Potassium acetate Rubidium chloride pH 5,8 at 25 °C	10 mM 15 % (v/v) 30 mM 100 mM

TFB II	MOPS	10 mM
	Rubidium chloride	10 mM
	Calcium chloride	75 mM
	Glycerol	15 % (v/v)
	pH 7,0 at 25°C	
RNA storage and handling solution	Trizol	
Homogenization buffer	Natriumchloride	100 mM
	Magnesiumchloride	2 mM
	HEPES	25 mM
	EDTA	0.1 mM
	Igepal CA-630	0.1 % (v/v)
	pH 7.5 at 25°C	
M9 buffer	Natriumchloride	85 mM
	KH_2PO_4	22 mM
	Na_2HPO_4	22 mM
	$MgSO_4$	1 mM
Washing solution	Penicillin	100 µg/mL
	Streptomycin	100 units/mL
	in 1x Hanks Balanced Salt Solution	

2.2.4 Culture Media and supplements

LB-medium	Lennox L Broth Base	20 g
	A. dest.	ad 1 L
	autoclave	
LB-Agar	Lennox Broth	20 g
	Bacto Agar	15 g
	A. dest.	ad 1 L
	autoclave	
SOC-regeneration medium	Invitrogen	
Worm culture medium	RPMI-1640	
	Penicillin	100 U/mL
	Streptomycin	100 µg/mL
	HEPES	10 mM
Cycloheximide inhibitory solution	Cycloheximide	70 mM
	in RPMI-1640	

2.2.5 Plasmids

For sequencing PCR fragments were cloned into the pGEM-T Easy vector (Promega) or optionally into the pCR4-TOPO vector (Invitrogen). For expression in *Escherichia coli* the pJC45 vector kindly supplied by PD Dr. Joachim Clos (BNI) was used.

2.2.6 Oligonucleotides

The oligonucleotides were designed independently and ordered at Operon (Cologne) at a concentration of 10 mmole.

Table 2.2.6-*1* Oligonucleotide sequences

Name	Sequence (5' - 3')
Spliced leader 1 (SL-1)	ggt tta att acc caa gtt tga g
T7I	gag aga gga tcc aag tac taa tac gac tca cta tag gga gat t_{24}
T7II	gag aga gga tcc aag tac taa tac gac tca cta tag g
SrGal2for1 (formerly SS00840for1)	tga tat tag aat tcg tgc tc
SrGal2for2 (formerly SS00840for2)	aca aat tta cca ttc ttg c
SrGal2rev1 (formerly SS00840rev1)	tta caa aag ttg gat tcc ag
SrGal2rev2 (formerly SS00840rev2)	tca aca tct cca cct att tg
SrGal2RTPCRf1	ttc ctc ttc ata ttt cta ttc g
SrGal2RTPCRr1	tgg taa taa gaa tat cac ctt c
SrGal22f3	tca aca aac tca tgt cat tgc aa
SrGal22f4	tta tac ttg acc ttt atg ctc aac
SrGal22VLKf1	atg gaa cca act gca ccc a
SrGal22VLKr1	cta gat att tga aat tgt aac taa
SrGal22VLK2f1	atg gaa cca act gca cc
SrGal22VLK2r1	cta gat att tga aat tgt aac t
SrGal22ECHind3f	aag ctt atg gaa cca act gca cc
SrGal22ECBamH1r	gga tcc cta gat att tga aat tgt aac t
SrGal22RTPCRf1	cac agc ctt aaa tgg tta tag
SrGal22RTPCRr1	tga gtt tgt tga gca taa agg
SrGal1f1	agt acc ata caa atc tca ac
SrGal1f2	ttc aag aga aat ttg aac c
SrGal1r1	tca gga gta gcg tag ata ag
SrGal1r2	aag ttg act ttt cta ctg gg
SrGal1ECHind3f	aag ctt atg gct gat gaa aaa aaa agt
SrGal1ECXho1r	ctc gag tta att aat ttg aat tcc agt
SrGal1RTPCRf1	cag tac cat aca aat ctc aac
SrGal1RTPCRr1	gtt cac cct tct taa ttg gat
SrGal3f1	atg tct act gaa act cat t
SrGal3f2	agt acc ata tcg ttc aaa ac

Table 2.2.6-1 continued

Name	Sequence (5' - 3')
SrGal3r1	ttc cag gaa cta acc cat ctc c
SrGal3r2	tca cca tca act gag aaa tg
SrGal3VLK2f1	atg tct act gaa act cat tta c
SrGal3VLK2r1	tta aac caa ctg aat tcc agt a
SrGal3RTPCRf1	taa tac att caa taa ggg aga at
SrGal3RTPCRr1	ggt tat ccc act ttc ata tg
SrGal3ECHind3f1	aag ctt atg tct act gaa act cat tta c
SrGal3ECBamH1r1	gga tcc tta aac caa ctg aat tcc agt a
SrGal11f1	gga tca cat ttt tca ata cat g
SrGal11f2	gaa gat aga cat cat aat cct t
SrGal11RTPCRf1	atg cat att att gac aac cct
SrGal11RTPCRr1	gga tta tga tgt cta tct tca
SrGal5f1	act aat agc att gat tgc tac
SrGal5f2	act tct gta ttt gga agt tc
SrGal5r1	tgg agt ttc aaa aag aat tcc
SrGal5r2	tga tca att cca tca att gag g
SrGal5RTPCRf1	caa atc ctt tca aag cta act
SrGal5RTPCRr1	tat caa tac gtt ttc ctc ttg
SrGal21f1	ggt gtt caa tta tat aat gtt tc
SrGal21f2	att ata atg tac cat atg aag ca
SrGal21r1	atg atc ata aac tcc aac cg
SrGal21r2	tac act aat aaa tca aaa gta cg
SrGal21RTPCRf1	ggt gtt caa tta tat aat gtt tc
SrGal21RTPCRr1	gtc gtc tgt cat tgt att ac
POPf1	tat gaa tat tta gaa aat tta caa gg
POPf2	aat ctt aat aaa ata tca aat aaa tat t
POPr1	taa ttt tat acc ttt ttt atg aaa tat
POPr2	att gga atc att gta cca tct tt
POPr1f	gta aag atg gta caa tga ttc caa tg
POPr2f	gga aat tta atg gaa atg aaa cat ggt
Astf2	tta tac atg aaa cat ctc atg ctc
Astr3	tta ttt aaa act ttt gaa ctt aat tg

2.2.7 Enzymes

Enzyme	Company/Origin	Description
DNase I	NEB	RNase-free DNase I
Proteinase K	Qiagen	Cystein-protease
Restriction enzymes	New England Biolabs, Fermentas, Roche	Restriction endonucleases Type II

Reverse transkriptase	Qiagen Invitrogen	SensiScript SuperScriptII, Super-ScriptIII
RNase A	Roche	DNase-free RNase
Taq-Polymerase	New England Biolabs Invitrogen	DNA-Polymerase DNA-Polymerase
T4-DNA-Ligase	New England Biolabs, Fermentas	DNA ligation
Trypsin	Promega	Tryptic digestion for MS

2.3 Methods

2.3.1 Working with *Strongyloides ratti*

2.3.1.1 Infection of hosts

For the infection freshly harvested iL3 were used in most cases. If no fresh iL3 were available, larvae that have been stored at 4°C in 1x phosphate buffered saline (PBS) or tap water were taken for the infection. Larvae were counted under the microscope. After counting, larvae were either diluted or concentrated depending on the density of the suspension to a final concentration of 10 iL3/µL. For the infection four week old male wistar rats were used. In the beginning the rats were infected subcutaneously with 1,800 iL3 using a syringe. The amount of iL3 was increased to 2,500 iL3 after the initial infections.

2.3.1.2 Charcoal culture and Baermann apparatus

For the collection of faecal pellets rats were placed on steel mesh from the fifth day after infection. To dispense rat urine, paper towels were placed under the steel mesh. Pellets were collected into crystallisation dishes every 24 hours. 1xPBS or water at room temperature were added to the pellets until they were almost covered with liquid. After one hour incubation about half of the volume of the charcoal crystallisation dish was filled with charcoal. To prevent absorption of water the charcoal was always kept at moisturized conditions. Everything was thoroughly mixed and the charcoal was stacked as shown in figure 2.3.1.2-*1* A. In case the mixture was too dry water or 1xPBS was added covering about 0.5 cm of the bottom. The dishes were covered and placed into an incubator at 26°C.

For the separation of the respective stages from the charcoal culture the Baermann method was used (Fig 2.3.1.2-*1* B) (Whitehead, 1965). Warm water (35–38°C) was filled in a funnel which was closed at the lower end using a clamp. A steel sieve was placed on top of the funnel. About ¼ of the steel sieve had to be covered with water. A piece of cotton was laid into the sieve and the charcoal culture was carefully transferred into the cotton. A lamp was placed directly next to the funnel to maintain warm temperatures.

Figure 2.3.1.2-*1* (A) Setup for a charcoal culture dish. (B) Setup of the Baermann funnel routinely used at the BNI for isolation of *S. ratti* stages from faecal cultures.

2.3.1.3   Preparation of iL3

For the isolation of iL3 faecal pellets were collected on days 6-16 after subcutaneous infection of male Wistar rats with 1,800–2,500 iL3. Charcoal coprocultures were set up and incubated at 26°C. The culture dishes were incubated 5–7 days for the collection of iL3. For the recovery of iL3 the Baermann method was used as described in the previous section. After separation of iL3 from the charcoal culture the larval suspension was pre-cleaned by rinsing the larvae a few times in a suction filter. It was avoided to completely dry the suction filter. Infective L3 were then washed four times in sterile washing solution.

2.3.1.4 Preparation of the free-living stages

For the isolation of free-living stages faecal pellets were collected on days 6-16 after subcutaneous infection of male Wistar rats with 1,800–2,500 iL3. Charcoal coprocultures were set up and incubated at 26°C for 24 hours. The following steps were performed as described in the previous section.

2.3.1.5 Preparation of parasitic females

For the recovery of parasitic females (pf) male Wistar rats were infected with 2,500 iL3. On days six and seven past infection the rats were sacrificed and the small intestine was removed beginning from the stomach and ending approximately 10 cm before the appendix. The intestine was pre-cleaned by emptying the contents through careful squeezing of the intestinal walls. After that the intestine was opened longitudinally using scissors and cut into strips of about 8–10 cm length. The strips were then washed three times by gentle shaking in three different glasses filled with 500 mL of water or PBS to remove residual debris. To separate the females from the tissue the strips were placed directly on the sieve without the cotton in a Baermann apparatus and incubated for three hours. After sedimentation of the females 50 mL of the solution in the Baermann funnel were filled in a Falcon tube. After a second sedimentation of the females they were transferred to a 1.5 mL tube using 1 mL pipet tips and washed at least six times in sterile washing solution. Between the washing steps the tube was centrifuged at 1,000 rpm for one minute. Using the repeated centrifugation steps at low speed led to a separation of parasitic females from tissue and residual eggs and first stage larvae. The female suspension was now used for *in vitro* culture, the preparation of extracts or for the preparation of RNA.

2.3.1.6 Preparation of E/S products

Freshly harvested and washed iL3, pf or free-living stages (fls) were carefully suspended in sterile worm culture medium under the laminar flow hood. The incubation densities were not exceeding 30,000 iL3/mL, 15,000 fls/mL and 100 pf/mL. Depending on the amounts either culture dishes or culture flasks were used. iL3 were incubated 24 hours and pf 72 hours at 37°C. Fls were incubated at 26°C for 24 hours. After the incubation period vitality and sterility were checked under the microscope. An additional test for sterility was performed by placing 5 µL of culture medium on blood agar plates and subsequently incubating the plates at 37°C for 24 hours. Only sterile cultures were used for further processing.

For the inhibition of protein synthesis sterile cycloheximide (CHX) and sodium azide stock solution was added directly before the incubation of iL3. CHX was added to a final concentrations of 50 and 70 µM. Sodium azide was added to final concentrations of 0.5 and 1.0%. After two hours of incubation the culture medium was removed, an equal amount of new medium was added and the larvae were further incubated for 24 hours.

0.5 M prolyl oligopeptidase inhibitor stock solutions of compounds 1A, 1B, 2A and 2B were prepared in PBS and sterile filtered. The stock solutions were added directly before the incubation of parasitic females at final concentrations between 1–10 mM.

2.3.1.7 Preparation of worm extracts

Freshly harvested worms were washed three times in M9 buffer and twice in homogenisation buffer (HB). After removal of the supernatant the worms were either directly processed further or frozen at -70°C until use. For further processing worms were placed on ice and HB buffer with 2 mM dithiothreitol (DTT), 1x Complete Mini protease inhibitor was added. One steel bead per tube was added and the worms were vortexed for ten minutes, after removal of the bead, sonicated for 30 seconds on ice four times. After spinning at 14,000 g for 20 minutes the supernatant was transferred into a new tube.

2.3.1.8 Whole worm analysis

Frozen larvae were picked individually using a 10–100 µL pipet under the microscope. Picked individuals were transferred to a clean tube and centrifuged at maximum speed for two minutes. The aqueous phase was discarded carefully using a 10 µL pipet without accidentally taking out any worms. 20 µL lithium dodecyl sulphate (LDS) loading buffer was added to each sample and heated to 95°C for ten minutes. The buffer then was loaded on NuPage®Novex®tris acetate mini gels and the electrophoresis performed until the buffer front had moved approximately 2 cm into the gel. The gel was Coomassie stained. After destaining each lane was cut out in either one or more gel strips depending on the number of bands visible. The samples then were reduced, alkylated, subjected to in-gel tryptic digest and loaded onto a mass spectrometer. Database searches were performed as described in 2.3.4.

2.3.2 Molecular biological methods

2.3.2.1 *E. coli* culture

E. coli DH5α cells were cultured aerobically in Lennox Broth (LB)-medium supplemented with ampicillin at 37°C and shaking at 200 rpm using a shaking incubator. The amount of cells was determined photometric at a wavelength of $\lambda = 600$ nm (OD_{600}). An $OD_{600} = 1$ equals approximately 2×10^9 cells. Ampicillin was added as a selection marker when working with transfected bacteria. For stock solutions 1/5 vol. glycerine was added after the cells reached the exponential growth rate. The stock solutions were stored at -70°C.

2.3.2.2 Generation of competent bacteria and transformation

In order to generate competent bacteria able to take up foreign DNA molecules with great efficiency they were treated with high concentrations of rubidium chloride. For this purpose 400 mL *E. coli* cultures were grown at 37°C in LB-medium until they reached an OD_{600nm} of 0.6 and subsequently cooled on ice. The cells then were centrifuged for ten minutes at 3,000 x*g* at 4°C. The supernatant was discarded and the cells were resuspended in 30 mL ice cold TFB I and centrifuged again. The supernatant was discarded, the cells were resuspended in 2 mL ice cold TFB II and divided into aliquots of 100 µL. Before storage at -70°C the cells where frozen in liquid nitrogen.

For transformation 10 µL from the ligation step were added to 100 µL of competent cells and placed on ice for 30 minutes. The cells then were heated at 42°C in a water bath for 90 seconds, cooled down on ice and supplemented with 250 µL SOC-medium. After one hour incubation at 37°C and 400 rpm on a shaking incubator the cells were plated on LB-Agar plates supplemented with ampicillin and incubated over night at 37°C.

Additionally OneShot™ TOP10 cells were purchased. The transformation was performed according to the manufacturer's protocol.

2.3.2.3 Plasmid preparations

For the preparation of plasmid DNA from bacteria the QIAprep Spin Miniprep Kit was used according to the manufacturer's protocol.

2.3.2.4 Expression of *S. ratti* galectin-3 in *E. coli*

For the prokaryotic expression the expression vector pJC45 was used. Primers containing restriction sites for the expression vector were designed. The 5'-end primer contained the palindrome *aagctt* which is the recognition site for *Hind*III. The 3'-end primer contained the palindrome *ggatcc* which is the recognition site for *Bam*HI. The PCR was performed using iL3 cDNA as a template. The resulting PCR product was cloned into the pGEM T easy vector. After purification of the resulting plasmid preparation the galectin-3 gene was excised using the respective restriction enzymes. The resulting fragments were separated on an agarose gel and the visible band at 850 base pairs was eluted using the Illustra™ GFX™ PCR DNA and Gel Band Purification Kit according to the manufacturer's protocol. The expression vector was linearised using *Hind*III and *Bam*HI and the purified galectin-3 gene was ligated into the vector using as described in section 2.3.2.13. The resulting vector pJC45/gal3 vector was sequenced to prove that gene was inserted into the correct reading frame. Competent bacteria, *E. coli* strain TOP10, were then transformed with plasmid DNA and plated onto LB plates containing appropriate antibiotics. After overnight cultivating, positive colonies were transferred to 25 mL LB medium with 100 µg/mL ampicillin and incubated overnight at 37°C under constant agitation. The non-induced overnight culture was transferred to a 1L LB expression medium and cultivated at 37°C under agitation until the culture reached an optical density (OD_{600}) of 0.6. The protein synthesis was induced by adding 1 M isopropyl-D-thiogalactopyranoside (IPTG). The protein expression was controlled by collecting small aliquots of the culture after IPTG induction every hour. After four hours of growth at 37°C, the bacteria were harvested by centrifugation at 4,000 xg and 4°C for 30 minutes. The bacterial pellet was stored at -20°C until further processing.

2.3.2.5 Purification of recombinant proteins by affinity chromatography

The recombinant protein contained an N-terminal histidine tag of ten amino acids. It was purified by nickel chelate affinity chromatography. For the purification the bacterial pellets from 3 L culture medium were combined and resuspended in 50 mL buffer B that contained urea for denaturation. The suspension was placed on a roller for two hours. Residual debris was pelleted by centrifugation (15,000 xg, 45 minutes, 4°C) and the supernatant was used for the purification procedure. For affinity chromatography a Ni-NTA agarose column (Qiagen) was loaded according to the manufacturer's protocol. The purified protein was then washed twice in 16 mL buffer C (pH 6.3) and eluted four times in 2 mL buffer D (pH 5.9), four times in 2 mL buffer E (pH 4.5). The eluted fractions were analysed by 1-D SDS PAGE and positive fractions were pooled. The pooled fractions were dialysed overnight in 2 M urea buffer (pH 8.0). The protein was then concentrated to a final concentration of 600 µg/mL using ultracentrifuge columns. During the centrifugation procedure the buffer was changed to 1xPBS. The protein concentration was measured using the Bradford assay.

2.3.2.6 Total RNA isolation

Fresh worms or larvae were washed 4x in washing solution, suspended in 1 mL Trizol solution and transferred into a 2 mL Eppendorf tube. They then were either directly processed or stored at -70°C for later processing. When processed one steel bead was added per tube and, to reach a maximum degree of soft disintegration, the tube then was vortexed for ten minutes on maximum speed. Following disruption 200 µL chloroform was added, kept at room temperature for 3 minutes and centrifuged for 15 minutes at 4°C/10,000 x g. The resulting upper phase was transferred to a new 1.5 mL Eppendorf tube, 500 µL isopropanol added and kept at room temperature for ten minutes. Then the preparation was centrifuged for ten minutes at 4°C/10,000 x g. After removal of the solvent by pipetting, the RNA pellet that formed during the centrifugation was washed by adding of 1 mL 75% ice-cold ethanol and brief vortexing followed by an additional centrifugation step for five minutes at 4°C/10,000 x g. The ethanol was discarded by pipetting and residual solvent evaporated from the open tube after 2 minutes drying at 37°C on the heating block. Finally the tube containing the RNA was directly placed on ice and 0.5 mL ethanol was added for storage at -70°C until further processing.

To retrieve better results when performing the reverse transcription it proved useful to reverse transcribe directly after RNA isolation. Then, instead of ethanol 16 µL DEPC water was added.

2.3.2.7 Reverse transcription

Before reverse transcribing of RNA it was incubated with RNase free DNase I (New Englans Biolabs, NEB) to exclude DNA contamination. DNase I reaction buffer was added to the equivalent of 10 µg RNA to a final volume of 100 µL. 2 µL DNase I was added and the reaction was incubated at 37°C for ten minutes. Then 1 µL 0.5 M EDTA was added and the enzyme was heat inactivated for ten minutes at 75°C. The following reverse transcription was performed either with the ProtoScript® First Strand cDNA Synthesis Kit (NEB) or using the SuperScript™ III Reverse Transcriptase (Invitrogen). In both cases the manufacturer's instructions were generally followed. Only the antisense primer was changed to the T7I primer as shown in table 2.2.6-*1*.

2.3.2.8 Polymerase chain reaction

The principle of the Polymerase Chain Reaction (PCR) is used for the amplification of DNA sequences and has been described by Sambrook (1989). In this work all PCRs were performed using the following components:

10x buffer (NEB)	5 µL
Forward primer (20 pmol)	1 µL
Reverse primer (20 pmol)	1 µL
GeneAmp®dNTP mix (AB)	1 µL
Taq polymerase (NEB)	0.3 µL
DNA	0.1-0.5 µg
HPLC-H$_2$O ad	50 µl

The denaturation step was set to five minutes at 95°C and the elongation was performed at 72°C between 30–120 seconds depending on the fragment size to be amplified. The annealing temperature is related to the length and specificity of the primers used and may not exceed the melting temperature (Tm) which was calculated using the following formula: Tm = (A+T) x 2 + (G+C) x 4. The annealing time was set to 40 seconds and 30–35 cycles were repeated. The result of the amplification was tested on an agarose gel.

2.3.2.9 Purification of DNA fragments

For the purification of PCR-products and DNA-fragments the Invisorb DNA Extraction Kit (Invisorb, Berlin) was used. The DNA band of interest was cut from an agarose gel at a wavelength of 366 nm and transferred to a 1.5 mL tube. The elution was performed according to the manufacturer's protocol. The quality of the DNA was subsequently tested on an agarose gel and the concentration was measured.

2.3.2.10 5'- and 3'-cDNAs amplification

To obtain 3'-cDNA end, 3'-Rapid Amplification of cDNA Ends (RACE) was performed. The 3'-RACE is a method that generates full-length cDNAs by utilising 3'-oligo-dT-containing primer complementary to the poly(A) tail of mRNA at the first strand cDNA synthesis. 5 µg total RNA was reverse-transcribed using Superscript III and GeneRacer (Invitrogen) oligo dT primer according to the manufacturer's instructions. RACE fragments were then isolated by RT-PCR using Taq polymerase, gene-specific forward primers and the T7II primer as reverse primer. The DNA fragments were cloned into pGemTeasy vector and sequenced.

5' cDNA fragments were obtained in different ways. The first approach was to perform Spliced Leader (SL) RT-PCR. 5 µg total RNA was reverse-transcribed using Superscript III and a gene specific primer according to the manufacturer´s instructions. RT-PCR was performed using Taq polymerase (Invitrogen), SL forward primers and the gene specific reverse primers. In case the gene did not carry the spliced leader sequence 5' RACE was performed according to the manufacturer's protocol. Resulting DNA fragments were cloned into pGemTeasy vector and sequenced.

2.3.2.11 Agarose gel electrophoresis

Agarose gel electrophoresis was performed as previously described (Sambrook, 1989) using TAE buffer and agarose concentrations between 0.7-1.5% (w/v) depending on the size of the fragments to be separated. 5 µL ethidiumbromide per 100 mL agarose solution was added before pouring the gel. The voltage was set to 100 V. DNA bands were visualised using a UV-transilluminator and immediately photographed.

2.3.2.12 Determination of nucleic acid concentrations

The determination of DNA and RNA was performed using a photometer by measurement of the absorption at 260 and 280 nm. The nucleotides were always diluted 1:99 in silica cuvettes.

2.3.2.13 Ligating of DNA

PCR products were either ligated into the pGEM-T Easy vector (Promega) for sequencing or into the pJC45 vector for prokaryotic expression. The ligation reaction for pGEM-T Easy was set up with 3 µL PCR product according to the manufacturer's protocol and incubated for two hours at room temperature.

For the ligation reaction with pJC45 two reactions with different insert/vector ratios were performed as follows:

	Reaction 1	Reaction 2
10 x T4 ligase buffer (NEB)	1 µL	1 µL
Linearised pJC45	3 µL	2 µL
Insert	1 µL	2 µL
T4 DNA ligase (NEB)	1 µL	1 µL
Deionised water	4 µL	4 µL

The ligations were incubated two hours at room temperature. The resulting construct was cloned into TOP10 cells for expression as described in section 2.3.2.4.

2.3.2.14 Restriction analysis

Restriction enzyme digestions were performed by incubating double-stranded DNA molecules with an appropriate amount of restriction enzyme(s), the respective buffer as recommended by the supplier(s), and at the optimal temperature for the specific enzyme(s). In general 20 µL digest were planned. For preparative restriction digests, the reaction volume was scaled up to 50 µL. Digestions were composed of DNA, 1 x restriction buffer, the appropriate number of units of the respective enzyme(s) (due to the glycerol content, the volume of the enzyme(s) added should not exceed 1/10 of the digested volume), and the sufficient nuclease free water to bring the mix to the calculated volume. After incubation at the optimal temperature for a reasonable time period (2-3 hours or overnight), digests were stopped by incubation for 20 minutes at 65°C. If reaction conditions of enzymes were incompatible with each other, DNA was digested successively with the individual enzymes. Between individual reactions, DNA was purified (see section 2.3.2.9 DNA fragment purification).

2.3.2.15 DNA sequencing

All PCR products of interest were cloned into the vectors mentioned in section 2.2.5 and sent to Agowa (Berlin) for commercial sequencing using the M13 forward primer or gene specific primers.

2.3.3 Biochemical methods

2.3.3.1 Determination of protein concentration by Bradford assay

The absorbance maximum for an acidic solution of Coomassie Brilliant Blue G-250 dye shifts from 465 nm to 595 nm when binding to protein occurs. The concentrations of protein solutions can be determined through comparison to a standard curve. All protein concentration measurements were performed using the Bio-Rad Protein Assay according to the manufacturer's protocol.

2.3.3.2 SDS-polyacrylamide gel electrophoresis (PAGE)

The separation of proteins using SDS PAGE was performed using the discontinuous system as described in the literature (Laemmli, 1970) which is based on the principle that proteins can be separated depending on their molecular mass under denaturing conditions. For all gels performed at the BNI Bio-Rad gel chambers were used. The gels were 8.3 x 7.3 cm in size. The gel concentrations were between 10% and 15% depending on the size of the proteins to be separated. The composition of the gels is described in table 2.3.3.2-1.

Table 2.3.3.2-1 Composition of SDS gels.

	Separation gels		Stacking gels
Acrylamide concentration (%)	10	15	4
4 x gel buffer (mL)	1	1	0.5
Acrylamide solution (mL)	1.33	1.99	0.26
Distilled water (mL)	1.67	1.01	1.24
APS (10% w/v) (µL)	30	30	15
TEMED (µL)	10	10	6

2.3.3.3 Coomassie staining of polyacrylamide gels

After SDS PAGE separation, the gel was washed three times in 2.5% Triton X-100 and stained with 0.1% (v/v) Coomassie brilliant blue, Roti®-Blue (Carl Roth). After destaining (Coomassie destaining solution), either protein bands on regular SDS PAGE gels or proteolytic degradation of the gelatin substrate was visible as unstained bands on the blue gel.

2.3.3.4 Silver staining of polyacrylamide gels

For silver staining all steps were performed at room temperature. After SDS PAGE, gels were fixed with fixation solution I and II for ten minutes, respectively and then washed in distilled water for ten minutes. Afterwards the gels were swayed in staining solution IV for ten minutes and finally in the developer solution until bands became visible. The gels were stored in distilled water until drying.

2.3.3.5 Substrate gel electrophoresis - zymogram

A detection of proteases is possible using the substrate gel electrophoresis in which a protein is added to the gel during the preparation (McKerrow, 1985). In all substrate gel experiments gelatin was added as substrate. Excretory/secretory products and extracts were separated under non-reducing conditions. SDS is inhibiting the activity of enzymes. Therefore the gels were washed three times in Triton-X-100 for 20 minutes to remove residual SDS. Afterwards the gels were washed three times in distilled water for five minutes each and finally incubated for 15 minutes in renaturation buffer. The gels then was placed in new renaturation buffer and incubated overnight at 37°C. Proteolytic bands are visualised using Coomassie staining. They appear colourless against a blue background.

2.3.3.6 Gelatin gel overlay

The direct determination of the relative molecular weight of proteases in the substrate gel electrophoresis with molecular weight markers is not possible. The migration behaviour of the proteins is influenced by the polyacrylamide-gelatin matrix of the substrate gel. To estimate the relative molecular weight of the protease the samples and a molecular weight marker were separated with SDS PAGE without the addition of the substrate gelatin under non-reducing conditions. The SDS PAGE gel was subsequently washed ten minutes in 0.1 M Tris-HCl buffer (pH 7.5). An additional gel containing 10% gelatin was prepared without SDS and stacking gel. Both gels were then laid on top of each other and fixed tightly using the glass plates and the gel frame of the BioRad system. The stacked gels were placed in tris buffer and incubated one hour at 37°C. The gels were then separated and the gelatin gel was incubated in tris buffer at 37°C overnight and stained with Coomassie. The SDS PAGE gel was stained using Coomassie or silver staining. All gels were dried and by placing them on top of each other the molecular weight of the protease was determined.

2.3.3.7 Lactose affinity separation

Proteins were diluted with 4mM 2-mercaptoethanol-PBS, pH 7.0 (MePBS) to final concentrations of 200 µg/mL total protein content. 500 µL protein preparation was added to 100 µL lactosyl sepharose beads and allowed to react at room temperature for two hours while shaking. After sedimentation of the beads the supernatant was collected and marked as unbound fraction. The beads were then washed six times with an equal volume of MePBS. The bound components were eluted incubation in an equal volume of 10% lactose in MePBS for two hours at room temperature. After sedimentation of the beads the supernatant was collected and marked as eluate one. The elution step was repeated and the supernatant was marked as eluate two. Finally an equal volume of SDS loading buffer was added and the beads were incubated for five minutes at 95°C. After sedimentation of the beads the supernatant was marked as eluate three. The unbound fraction and eluates one and two were concentrated on 10 kD Amicon Ultra spin filters and all samples were loaded onto an SDS gel. PAGE was performed and the gel stained in Coomassie staining solution.

2.3.3.8 One dimensional-electrophoresis and band excision

All samples were reduced by adding 10 mM DTT and incubating at 56°C for 45 minutes. For alkylation 55 mM iodoacetamide was added and the samples were incubated 30 minutes at room temperature in the dark. Then, NuPAGE® LDS 4x sample buffer (Invitrogen) was added and the samples were loaded on 12% NuPAGE® Novex® acetate mini gels. Gels were stained with SimplyBlue™ safe stain. Gel bands were then cut from the entire lanes with a scalpel.

2.3.3.9 Tryptic digestion

Prior to tryptic digestion gel pieces were washed five minutes in 0.1 M ammonium bicarbonate, destained in ammonim bicarbonate containing 1/3 acetonitrile for 60 minutes and finally dehydrated in 100% acetonitrile.

For in-gel tryptic digestion the gel pieces were hydrated 45 minutes in sequencing grade trypsin solution (Promega) on ice. The remaining trypsin solution was discarded, gel pieces were covered with ammonium bicarbonate and incubated overnight at 37°C. The gel pieces then were washed in ammonium bicarbonate containing 1/3 acetonitrile and the supernatant was transferred into a new 1.5 mL tube. Remaining peptides were extracted with 100% acetonitrile and the resulting solution was added to the supernatant from the washing step.

2.3.3.10 Liquid chromatography - tandem mass spectrometry (LC-MS/MS)

Peptides derived from in-gel digested proteins were analysed by online microscale capillary reversed-phase HPLC hyphenated to a linear ion trap mass spectrometer (LTQ, Thermo Scientific). Samples were loaded onto an in-house packed 100 µm i.d. × 15 cm C18 column (Magic C18, 5 µm, 200Å, Michrom Bioresource) and separated at approx. 500 nL/min with 30 min linear gradients from 5-40% acetonitrile in 0.4% formic acid. After each survey spectrum the six most intense ions per cycle were selected for fragmentation/sequencing.

Since LC-MS/MS was a new method utilised for protein identification in our laboratory some basic principles are briefly presented in the following. Figure 2.3.3.10-*1* shows the LTQ mass spectrometer and the additional components that were used for most of the experiments.

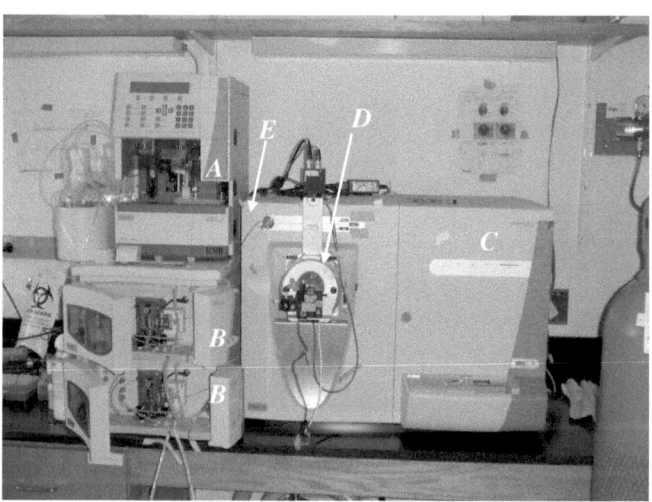

Figure 2.3.3.10-*1* Setup of the LTQ MS. **A**: autosampler, **B**: surveyor pumps, **C**: LTQ MS, **D**: ESI chamber, **E**: HPLC column

The peptides eluting from the HPLC column are transferred into the gas phase as peptide ions by electrospray ionization (ESI). The spray needle and the principle are shown and explained in figure 2.3.3.10-*2*.

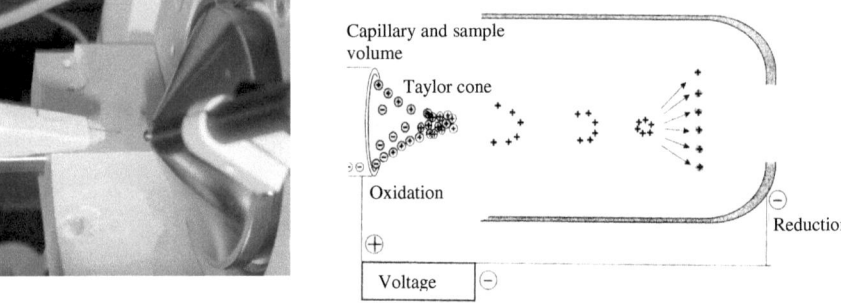

Figure 2.3.3.10-2 The formation of peptide ions by ESI. Left side: The spray needle and the entrance to the LTQ. Right side: The transfer of peptides to the gas phase (modified from Rehm, 2006). The pressure in the capillary and the voltage between the capillary and the counter electrode disperses the eluting solution. The droplets sizes decrease as the solvent disperses. At the same time the charge density increases, resulting in dispersion of the droplets due to electrostatic repulsion. Dehydrated and positively charged peptide ions emerge.

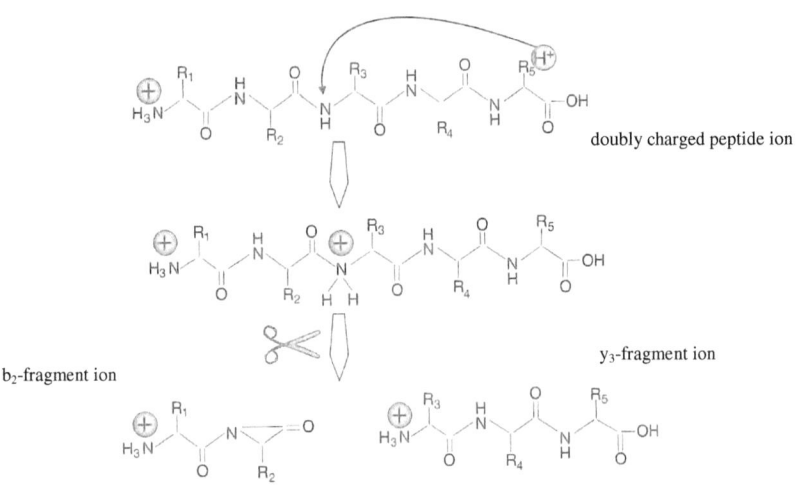

Figure 2.3.3.10-3 Fragmentation mechanism of a doubly charged peptide ion (modified from Rehm, 2006).

The resulting peptide ions then enter the linear triple quadrupole and finally reach the collision cell were they are further fragmented. Usually the peptides are fragmented at their weakest bonds, the covalent amide bonds, resulting in N- and C-terminal fragments. Unlike in the regular hydrolysis the amide bond cannot take up a water molecule during the fragmentation because the peptides are in the gas phase. Therefore the peptide ions are fragmented into b- and/or y-ions. The fragmentation mechanism of a doubly charged peptide ion is shown in figure 2.3.3.10-*3*. The resulting b- and y-ions are then separated according to their mass, and the separated ions are accelerated towards a detector where they are counted. The compiled spectra show the mass distribution of the ions produced from the sample.

The MS data were processed Xcalibur® mass spectrometry data system (Thermo Electron). The qualitative data can be reviewed using the Xcalibur® Qual Browser. Figure 2.3.3.10-*4* exemplary shows a Total Ion Current (TIC) chromatogram in the upper half and a mass chromatogram

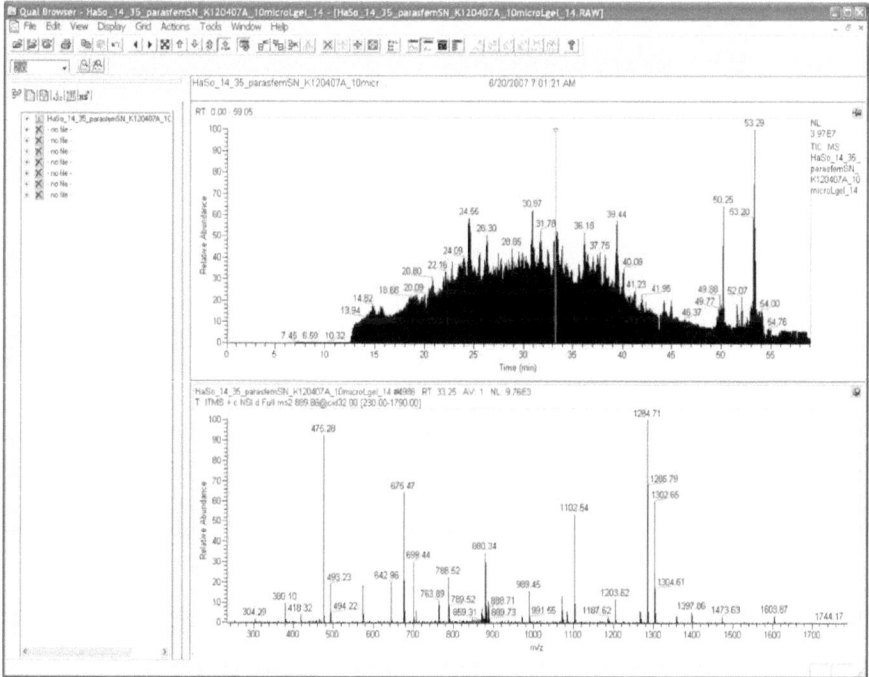

Figure 2.3.3.10-*4* Xcalibur® Qual Browser window. Upper half: Total ion current chromatogram. Lower Half: Mass chromatogram of a S. ratti protein sample

of a *S. ratti* protein sample in the lower half. A TIC chromatogram represents the summed intensities of all the ions in each spectrum plotted against chromatographic retention time. Each peak in the TIC represents an eluting compound, which can be identified from the mass scans recorded during its elution. A mass chromatogram shows the ion intensities of selected mass-to-charge ratios (*m/z*).

Prior to searching the spectra, the raw data files were converted into mascot generig format (mgf). The database searches were performed using the Applied Biosystems ProteinPilot™ search engine. The resulting data can be viewed in the software's user interface as shown in figure 2.3.3.10-5. The figure exemplarily shows the protein ID tab with the detected proteins in the upper pane, the protein groups in the middle and the sequence coverage in the lower pane.

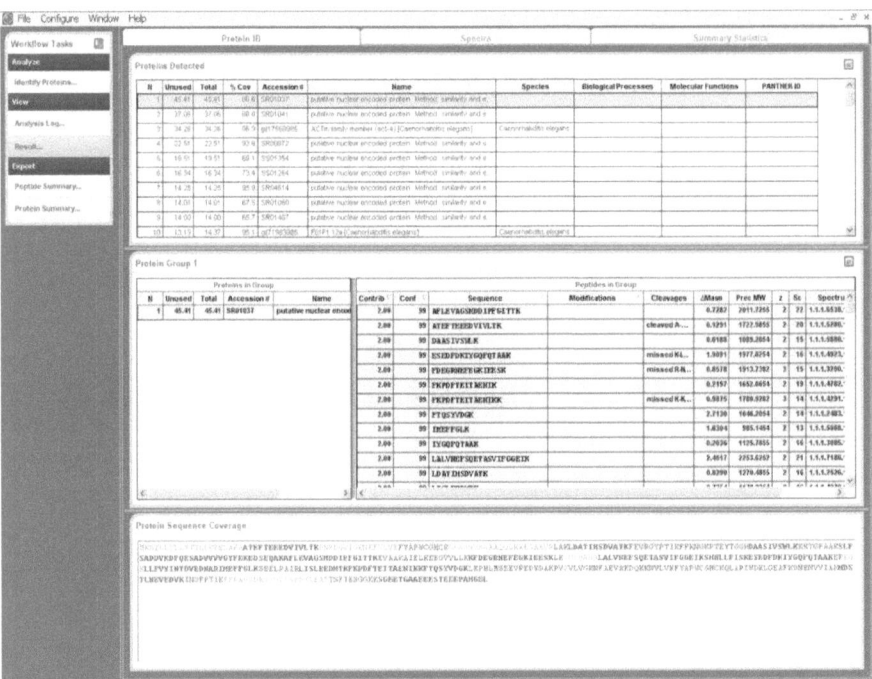

Figure 2.3.3.10-5 The Protein Pilot™ user interface.

Animals, Materials and Methods

By allocating the data files to their respective samples the protein composition of a protein band can be determined as exemplary shown in figure 2.3.3.10-6. The figure shows a gel slide prepared from iL3 E/S products.

Proteins Detected

N	Unused	Total	% Cov	Accession #	Name	Species
4	16.06	16.06	70.5	SR01037	putative nuclear encoded protein Method: similarity and e...	
5	14.10	14.10	85.8	SR02686	putative nuclear encoded protein Method: similarity and e...	
6	13.94	13.94	86.1	SR00900	putative nuclear encoded protein Method: similarity and e...	
7	12.36	12.36	63.4	SS01354	putative nuclear encoded protein Method: similarity and e...	
8	11.57	11.57	93.3	SR03119	putative nuclear encoded protein Method: similarity and e...	
9	11.03	11.03	83.9	SR00886	putative nuclear encoded protein Method: similarity and e...	
10	10.97	10.97	98.2	SR00386	putative nuclear encoded protein Method: ESTScan	
11	10.95	10.95	69.6	SR00920	putative mitochondrial protein Method: similarity and exte...	
12	10.69	10.69	70.8	SS03469	putative nuclear encoded protein Method: similarity and e...	
13	10.45	12.45	83.4	SR00627	putative nuclear encoded protein Method: similarity and e...	
14	10.39	10.39	95.3	SR00487	putative nuclear encoded protein Method: similarity and e...	
15	10.28	10.28	53.2	SS01564	putative nuclear encoded protein Method: similarity and e...	
16	10.10	10.10	64.0	SR01042	putative nuclear encoded protein Method: similarity and e...	

Protein Group 13

N	Unused	Total	Accessio...	Name	Contrib	Conf	Sequence
13	10.45	12.45	SR00627	putative nuclear encod	2.00	99	FEPGQTLIVK
2	19.30	19.30	HannsSS008	HannsSS00840	2.00	99	IVLNTFSNGDWGK

Figure 2.3.3.10-6 Protein composition of a single gel band.

Table 2.3.3-1 Mass shift of fragment ions from homologous peptides. The example shows tryptic peptides from two *S. ratti* galectins.

Residue	b (m/z)	y (m/z)	Residue	b (m/z)	y (m/z)
F	147.18	1132.35	F	147.18	1120.30
E	276.30	984.17	E	276.30	973.12
P	373.42	856.05	P	373.42	844.00
G	430.47	758.93	G	430.47	746.88
Q	558.60	701.88	Q	558.60	689.83
T	659.71	573.75	T	659.71	561.70
L	772.87	472.64	L	772.87	460.59
I	886.03	359.48	T	873.98	347.43
V	985.16	246.32	V	973.11	246.32
K	1113.34	147.19	K	1101.29	147.19

The protein group shows the EST cluster number SR00627 which is one of the galectin sequences presented in section 3.4.1. The use of mass spectrometry allows the differentiation between proteins showing a high homology like the *S. ratti* galectin family. The fragmentation of peptides differing in one residue leads to a mass shift in contiguous fragments from the same peptide as exemplary shown in table 2.3.3-*1*.

2.3.4 Bioinformatic procedures

2.3.4.1 Database searches

The data were searched against a custom database consisting of EST data sets from *S. ratti* and *S. stercoralis* and other sequence data available at NCBI (Tab 2.3.4.1-*1*).

Table 2.3.4.1-*1* Composition of the search database

Species	Number of sequences
C. elegans	22,977
S. ratti	5,237
S. stercoralis	3,479
Brugia malayi	1,907
Other nematodes	5,718
Total	39,318

Searches were performed using ProteinPilot^{TM} version 2.0.1 search engine (Applied Biosystems) in amino acid substitution mode. Following search parameters were selected:

- Sample Type: Identification
- Cys. Alkylation: Iodoacetamide
- Digestion: Trypsin
- Instrument: LTQ
- Special Factors: Gel-based ID
- ID Focus: Amino acid substitutions
- Search Effort: Thorough

Proteins were identified based on a minimum of 4.00 *unused protein score* equivalent to two or more unique peptides of 99% confidence. EST sequences were subjected to Basic Local Alignment Search Tool (BLAST) searches using NCBI BLAST. The resulting sequences were screened for signal sequences using SignalP 3.0.

2.3.4.2 Phylogenetic analysis

For the phylogenetic analysis of the galectin family *S. ratti*, *S. stercoralis* and other nematode galectin sequences were used. A datafile containing *S. ratti* full length and partial sequences and *S. stercoralis* EST sequences was prepared. Sequences from other nematodes were added from the NCBI protein database. The analysis itself was performed at the server of www.phylogeny.fr using the default settings. The server runs and connects various bioinformatic programs to reconstruct a robust phylogenetic tree from a set of sequences (Dereeper, 2008).

2.3.5 Immunological tests

2.3.5.1 Western Blot analysis

Western blot analysis was performed to detect the immunogenic properties of E/S products and extracts, to check the expression of galectin recombinant protein and to detect specific antibodies against galectin-3 in *S. ratti*-infected rats. The *S. ratti* iL3 E/S products and extracts, as well as the recombinant expressed proteins, were separated by SDS PAGE. After electrophoresis, the proteins were transferred onto a nitrocellulose membrane by using a mini-Trans Blot Cell in Bjerrum-Schafer-Nielsen transfer buffer for 50 min at 50 V. The membrane was blocked with 5% skim milk in PBS (w/v) overnight under constant agitation at 4°C. Subsequent to a wash step with TBS/0.05% tween 20 for ten minutes at RT, the nitrocellulose membrane was probed with a 1:20 dilution of sera from *S. ratti* infected Wistar rats. The unbound antibodies were removed by two wash steps with TBS/0.05% tween 20 and PBS for ten minutes respectively. The appropriate diluted (5% skim milk in PBS (w/v)) horseradish peroxidase-conjugated goat anti–rat IgG (Roche) secondary antibody was applied. The immunoreactive bands were visualised using 4-chloro-1-naphtol (Roche) colour development reagent according to the manufacturer´s protocol.

2.3.5.2 ELISA (Enzyme-linked immuno-sorbent assay)

Rat sera were analysed by ELISA for IgG antibodies to *S. ratti* antigens in E/S products, crude worm extracts and recombinantly expressed *S. ratti* galectin-3. 96-well microtiter plates (polystyrene microtiter plate, Maxi-Sorb, Nunc) were coated with 100 µl/well protein sample at a concentration of 200 ng/well in carbonate buffer (pH 9.6), sealed with Saran wrap and incubated overnight at 4°C. After removal of unbound protein by washing three times with 300 µl/well washing-buffer PBS/0.05% tween 20, the plate was blocked with 200 µL/well 5% BSA in PBS for one hour at 37°C. Different dilutions of rat sera were prepared in PBS/0.5% BSA prior to incubation at 37°C for one hour. Unspecifically bound proteins were removed by three washing steps. 100 µL/well 1:5,000 diluted anti-rat IgG peroxidase–conjugate antibody was added and incubated at RT for one hour followed by three washing steps. 100 µL/well peroxidase substrate solution was added for five minutes prior to 100 µL/well 1 M H_2SO_4 stop solution. For detection of antibody binding in all ELISA experiments, the optical density (OD) was measured at the absorbance at 450 nm (ref. 630 nm) using an ELISA reader (Dynatech).

3 Results

3.1 Establishing the *S. ratti* life cycle

The life cycle of *S. ratti* was established as a model system for the genetically related *S. stercoralis* infection (Viney, 1994). *Strongyloides* represents an ideal candidate to culture in the laboratory because unlike many other parasitic nematode life cycles it is neither difficult nor labour intensive to maintain in an environment that does not totally reflect their natural living conditions. The most practical laboratory hosts for the human parasite *S. stercoralis* have proven to be dogs, taking into account the large numbers of adult worms and higher output of faecal larvae. The use of such species, however, is both, prohibitively expensive and of ethical concern. Hence it was decided to study *S. ratti* in their natural host species, the rat, and establish the life cycle at the BNI in 2006 instead.

The iL3 needed for the initial infection of rats were supplied by the STI. For the maintenance of the *S. ratti* strain a four week infection cycle proved to be most feasible (Table 3.1-*1*) ensuring a secure perpetuation and at the same time minimising the number of rats needed for infection and the inevitable euthanisation. The collection period could not be prolonged because after 22–23 days the faecal egg output is reduced, leading to low iL3 recovery when performing the Baermann method. Also boosting the infection only shows a slight increase in egg output compared to the initial infection due to the development of a protective immune response. In case higher amounts of iL3 or different developmental stages are needed additional rats could be purchased and infected at any time during the four week infection cycle provided that enough iL3 were available. The time course of the infection was analysed and notably the progression consists of six phases. For the analysis of the faecal egg output it is adequate to measure the egg output by determining the number of iL3 that can be found in the faecal cultures after applying the Baermann method (Figure 3.1-*1*). In the first phase of about 4-5 days iL3 have to proceed to the intestine and develop to egg laying parasitic females. Thus, no iL3 could be detected in the Baermann funnel after coproculture. Two peaks of egg output could be observed beginning on days 5 and 11 (phase II and IV) and followed by a strong decrease on days 8 and 13 (phase III and V). At day 22 the number of larvae shows an additional decrease to a very low level and finally on days 29-30 no larvae or sometimes only a few larvae were found in the faecal pellets. The maximum yield of iL3, therefore, occurs at day 6–8 and 11–13 post infection.

Table 3.1-1 Scheme for the maintenance of the *S. ratti* life cycle at the BNI-Hamburg. At least two 4 week old male Wistar rats should be infected. Additional rats can be infected in case higher amounts of iL3 or other stages are needed. C - rats have to be kept on wood chippings, G - rats have to be placed on steel grids, **0-25** – Days post fection, CC – Charcoal Culture

Week	Day							
First week	M				Order rats	
	T	W	C		Arrival of rats	
	T	T	C	0	Infection of rats	
	F	C	1		
	S	C	2		
	S	C	3		
Second week	M	G	4		
	T	G	5	Collect pellets – CC1	
	W	G	6	Collect pellets – CC2	
	T	G	7	Collect pellets – CC3	
	F	C	8	Collect pellets – CC4	
	S	C	9		
	S	G	10		
Third week	M	G	11	Collect pellets – CC5	Baermann CC1
	T	G	12	Collect pellets – CC6	Baermann CC2
	W	G	13	Collect pellets – CC7	Baermann CC3
	T	G	14	Collect pellets – CC8	Baermann CC4
	F	C	15	Collect pellets – CC9	
	S	C	16		
	S	G	17		
Fourth week	M	G	18	Collect pellets – CC10	Baermann CC5/6
	T	G	19	Collect pellets – CC11	Baermann CC7
	W	G	20	Collect pellets – CC12	Baermann CC8
	T	G	21	Collect pellets – CC13	Baermann CC9
	F	C	22	Collect pellets – CC14	
	S	C	23		
	S	C	24		
Fifth week	M		25	Order new rats / Sacrifice old rats	Baermann CC10/11
	T				Baermann CC12
	W	C		Arrival of new rats	Baermann CC13
	T	C	0	Infection of new rats	Baermann CC14
	F	C	1		
	S	C	2		
	S	C	...		

Results

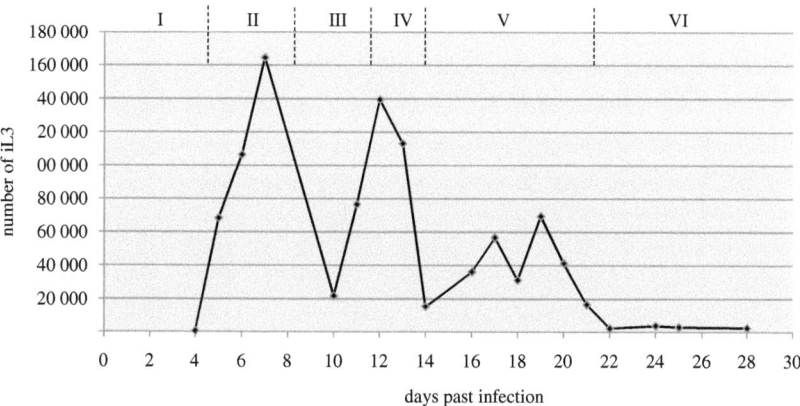

Figure 3.1-1 Time course of a *S. ratti* infection. The graph displays the number of iL3 that can be obtained from faecal pellets by Baermann method from one rat infected with 1,500 larvae. Numbers I-VI show the different phases during the infection. Incubation condition for charcoal culture: 24 h at 36°C.

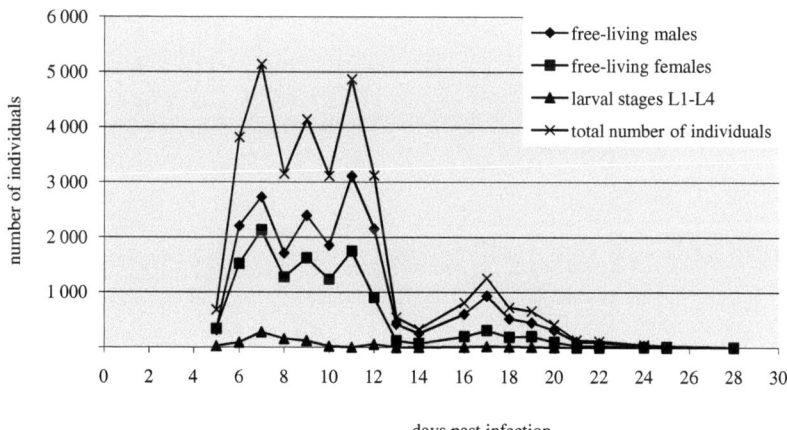

Figure 3.1-2 Development of different free-living stages during a *S. ratti* infection. The graph displays the number of free-living individuals that can be obtained from faecal pellets by Baermann method from one rat infected with 1,500 larvae. Incubation condition for cahrcoal culture: 24 h at 26°C.

A comparable profile results when the single free-living stages are counted after 24 hours of incubation at 26°C. Figure 3.1-2 shows the amounts of free-living males, females and the larval stages at a daily count during a 28 day period. After the fourth day there is a rapid increase in the numbers of free-living adults that decreases after day 12. On day 16 there is a slight increase in the number of males and females again. After day 20 only a few individuals can be found. The evaluation was stopped after day 28.

In addition the time periods for the development of the respective stages were measured (Figure 3.1-3). The cultures were examined at different time points during the culture at 26°C. All free-living stages showed a rapid development and were visible during the first 42 hours of culture. To get a 99% suspension of iL3 however took five days. For the parasitic phase it took about four days until a few eggs and first stage larvae were visible in the stool showing a strong increase on day five.

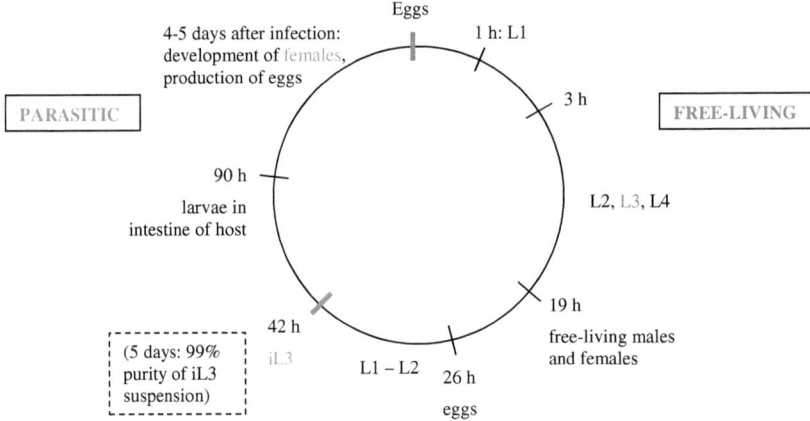

Figure 3.1-3 Rapid development of the life cycle. The free-living stages develop within 42 hours until first iL3 become visible. The parasitic phase is completed within four to five days. This figure does not represent the actual infection cycle in the laboratory. **h** – hours.

3.2 Optimising and processing of E/S products

Because of to the great yields of iL3 that can be achieved when performing the Baermann method, this stage was chosen for the evaluation of incubation conditions and the elaboration of the newly established protocols. Infective larvae were cultured under different conditions.

3.2.1 Dependence on the number of larvae

Increasing the number of incubated iL3 for 24 hours also showed an increase in protein production (Fig. 3.2.1-*1*).

The collagenase assay showed an increase of proteolytic activity. The protease was shown to be a zinc-dependent metalloprotease, belonging to the astacin family.

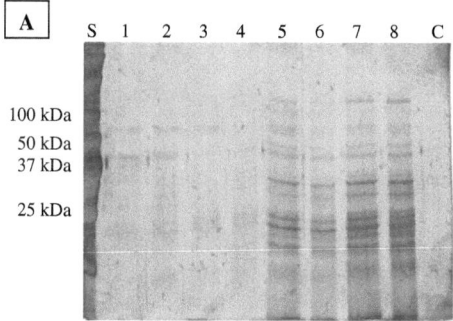

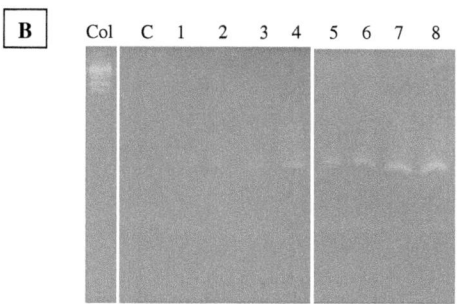

Lane	Number of incubated larve
C (Control)	0
1	5,000
2	10,000
3	20,000
4	40,000
5	60,000
6	80,000
7	100,000
8	150,000

Figure 3.2.1-*1* Amount of E/S products depends on number of infected larvae. (A) Silver staining of concentrated E/S products. (B) Zymogram of concentrated E/S products. (C) The table shows the number of iL3 per mL that were incubated for 24 hours. **Col** – Collagenase, **C** – Control, **S** – molecular marker

3.2.2 Detection of the metalloprotease

To show that the proteolytic activity is caused by a zinc-dependent metalloprotease, phenanthroline, a substance chelating zinc ions, was added. As a control a second gel was incubated in phenylmethanesulphonylfluoride (PMSF), an inhibitor of serine proteases (Fig 3.2.2-*1*). The gel shows the inhibition of the collagenase and the iL3 protease when incubated with phenanthroline. At the same time the serine protease trypsin is not inhibited. A second gel treated with PMSF shows the inhibition of trypsin only. The collagenase and the iL3 protease are still active.

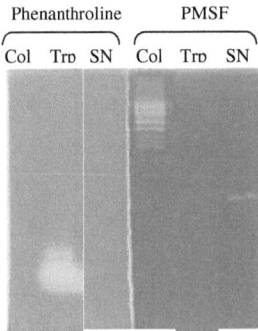

Figure 3.2.2-*1* Inhibition of the metalloprotease
left side: incubated with 20 mM phenanthroline, right side: incubated with 5 mM PMSF
Col – collagenase, **Trp** – trypsin, **SN** – iL3 culture supernatant

3.2.3 Size determination of the metalloprotease

The molecular size of the metalloprotease could not be determined in the gelatin gel owing to a delayed migration of proteins in the acrylamide-gelatin matrix. To determine the molecular weight of the secreted protease a gelatin overlay was performed. For this purpose an SDS PAGE without the addition of gelatin had to be performed. In the following step a gelatin gel was laid on the previously obtained SDS PAGE gel. During this step the proteins infiltrated the gelatin gel and the protease could unfold its activity. Finally all gels were stained and the gel band and the actual size of about 33 kDa could be determined (Fig 3.2.3-*1*).

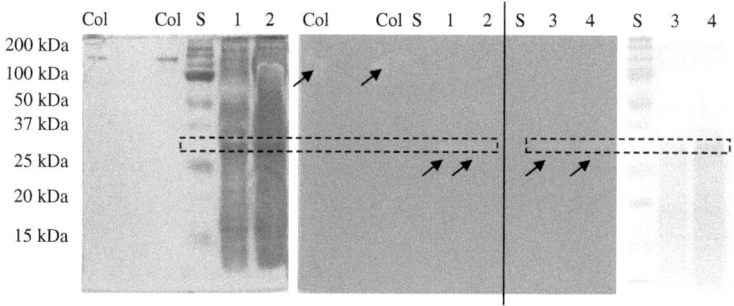

Figure 3.2.3-1 Gelatin overlay for size determination of the protease. The silver staining (right side) and the Coomassie staining (left side) represent the SDS PAGE gel used as template for the overlay (middle). **Col** – collagenase, **S** – molecular marker, **1, 2, 3** and **4** – sample iL3 supernatant. Arrows show the proteolytic activity

3.2.4 Dependence on the incubation times

Prolonging the incubation times from 24 hours to 48, 72 and 96 hours without a change of culture media only led to a slight increase of protein yield. Remarkably the different staining methods led to a significant shift in the band pattern (Fig 3.2.4-1). In the Coomassie stain significant bands can be seen at approximately 30 kDa and from 20 to 10 kDa whereas in the silver stain dominant bands can be seen below 20 kDa, between 30 and 37 kDa and between 50 and 150 kDa.

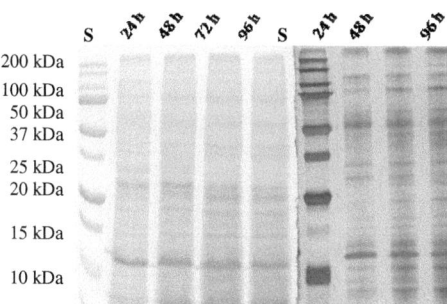

Figure 3.2.4-1 Coomassie and silver stain of iL3 E/S products. iL3 were incubated 24, 48, 72 and 96 hours. **S** – molecular marker

The zymogram shows that the above mentioned astacin metalloprotease is secreted only within the first 24 hours of incubation. A change of culture media and subsequent re-incubation leads to the production of further protein without proteolytic activity (Fig 3.2.4-2). Infective L3 were incubated 24 hours (lane 1) or 48 hours (lane 2) before the culture media was changed. In new culture medium the iL3 were then incubated another 24 hours (lanes 3 and 4). The band pattern in the silver stain shows similar significant bands as can be seen in figure 3.2.4-2. The zymogram shows proteolytic activity in the 24 and 48 hour incubation samples but no proteolytic activity in the samples that were subsequently incubated.

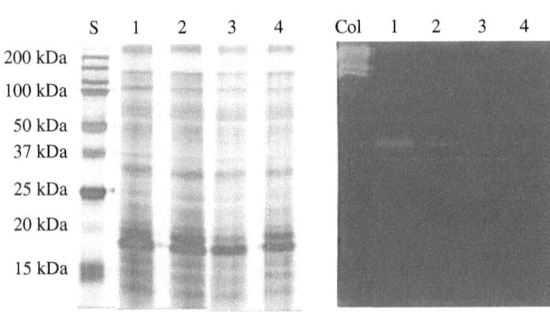

Figure 3.2.4-2 Silver stain (left side) and zymogram (right side) of iL3 E/S products incubated different time periods.
Samples - incubation times:
1 – 24 h
2 – 48 h
3 - 24 h (previously 24 h incubation and medium change)
4 – 24 h (previously 48 h incubation and medium change)
S – molecular marker, **Col** – collagenase
The size of marker of the silver stain does not represent the same size in the zymogram

3.2.5 Dependence on the incubation temperature

When the mobility of the larvae was inhibited, the protein synthesis was decreasing. The iL3 were incubated at 4°C and in parallel at 37°C. After 24 hours of incubation the culture media were changed and the incubation temperatures were changed from 4°C to 37°C and from 37°C to 4°C (Fig 3.2.5-1). The silver staining does not show any protein and the zymogram does not show any proteolytic activity in the samples that were incubated at 4°C (lanes 1 and 4). The samples that were incubated at 37°C show both protein and proteolytic activity. This observation is consistent with the movement of the iL3 at the respective temperatures. At 4°C the larvae were immobile whereas at 37°C the larvae started to move rapidly again.

Results

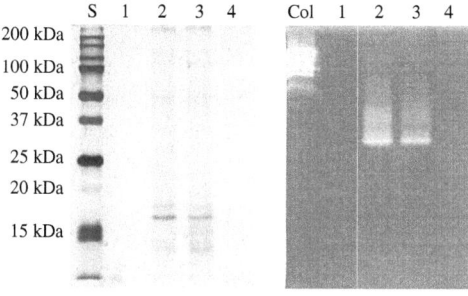

Figure 3.2.5-1 Silver stain (left side) and zymogram (right side) of iL3 E/S products incubated 24 hours at different temperatures.
Samples - incubation temperature:
1 – 4°C
2 – 37°C
3 – 37°C (previously at 4°C)
4 – 4°C (previously at 37°C)
S – molecular marker, **Col** – collagenase
The size of marker of the silver stain does not represent the same size in the zymogram

3.2.6 Inhibition of protein synthesis

To prove that the proteins visible in the gel are not proteins that are present due to leakage of dead or damaged individuals the iL3 were immobilised by the treatment at different temperatures or by the addition of substances directly to the culture medium. First the effect of heat and freezing was tested (Fig 3.2.6-1 A). Infective L3 were either heated to 70°C for ten minutes or put in liquid nitrogen for ten minutes. After both treatments no motile larvae were visible under the microscope. The larvae were then prepared for incubation in the same way as the untreated larvae and incubated for 24 hours. In the silver staining protein bands were strongly reduced after 70°C and less reduced after freezing in nitrogen.

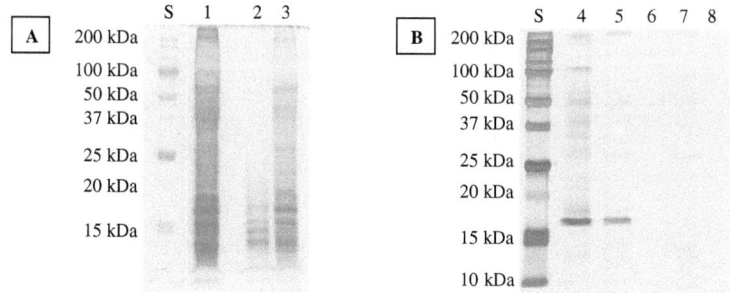

Figure 3.2.6-1 Different treatment methods for the incubation of iL3.
(A) Infective L3 were incubated untreated (lane 1), heated to 70°C for ten minutes (lane 2) and frozen in liquid nitrogen for ten minutes (lane 3) before the incubation. Protein is clearly visible in all samples.
(B) Infective L3 were immobilised using 50 and 70 mM cycloheximide (CHX) (lanes 5 and 6) and 0.5 and 1.0 % sodium azide (lanes 7 and 8). As a control one batch of larvae received no treatment (lane 4). The larvae were then incubated 24 hours. At the highest concentrations of CHX and sodium azide no protein is visible.

In addition, the secretion of protein was examined under the influence of physiological inhibitors, i.e. CHX, a substance inhibiting the protein synthesis of eukaryotes, and sodium azide. Figure 3.2.6-*1* B shows the treatment with 50 and 70 mM CHX and with 0.5 and 1.0 % sodium azide. As shown a decrease in protein synthesis was visible in the resulting gel. It was therefore decided to incubate the worms in 70 mM CHX as a control for the mass spectrometric analysis.

3.2.7 Differences in protein secretions and crude extracts

The secretions of *in vitro* cultured *S. ratti* stages were collected under serum free conditions. Owing to the large numbers of individuals needed to collect E/S products and at the same time visualise them on Coomassie-stained gels it was decided to restrain to the following stages: (1) iL3, (2) parasitic females (pf) and pooled free-living stages, consisting of the larval stages L1 – L4 and the free-living adult males and females (fls). To address the question whether the respective stages selectively secrete a specific subset of proteins, E/S products were compared to a soluble worm homogenate. Beside many matching bands 1-D gel electrophoresis also revealed that E/S products and crude extracts had some distinct band patterns (Fig 3.2.7-*1*).

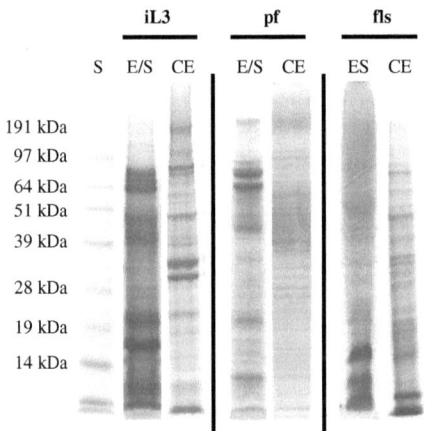

Fig 3.2.7-*1* 1-D SDS PAGE of iL3, pf and fls E/S products and extracts. The E/S products were concentratet 150-300 fold.
S – molecular marker, **E/S** – E/S products, **CE** – crude extracts

Results

3.2.8 Differences in protein secretion among various stages

As the aim of this work was to identify and describe E/S proteins which are abundantly or differentially produced by various developmental stages occurring in the life cycle, the three above mentioned stages were compared by 1-D SDS PAGE. Preliminary 1-D gel analysis of several different batches of concentrated E/S products (not shown) revealed a general consistency in protein composition, although minor differences were evident. To show that the E/S products are not simply reflecting non-specific leakage from the worms during the *in vitro* culture period iL3 were also incubated with the addition of CHX to the culture medium (iL3i) (Fig 3.2.8-*1*, also see Fig 3.2.6-*1*). Direct comparison of the excretions revealed distinct band patterns.

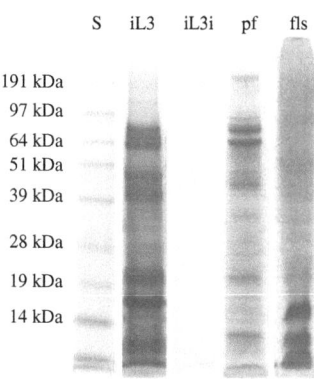

Fig 3.2.8-*1* 1-D SDS PAGE of iL3, pf and fls E/S products shows distinct band patterns of the respective stages.

3.2.9 Antibody recognition of E/S products

To define if the E/S proteins are target for immune recognition ELISA was performed. Rat sera were taken before subcutaneous infection with 1,500 iL3 and 42 days after infection. Additionally human sera of healthy Europeans and previously infected individuals living in endemic areas were tested.

The ELISA result shows the development of an IgG immune response against iL3 E/S products during the time of infection in rats (Figure 3.2.9-1). The sera from previously infected individuals with *S. stercoralis* showed a strong recognition of *S. ratti* iL3 E/S products compared to the human control serum.

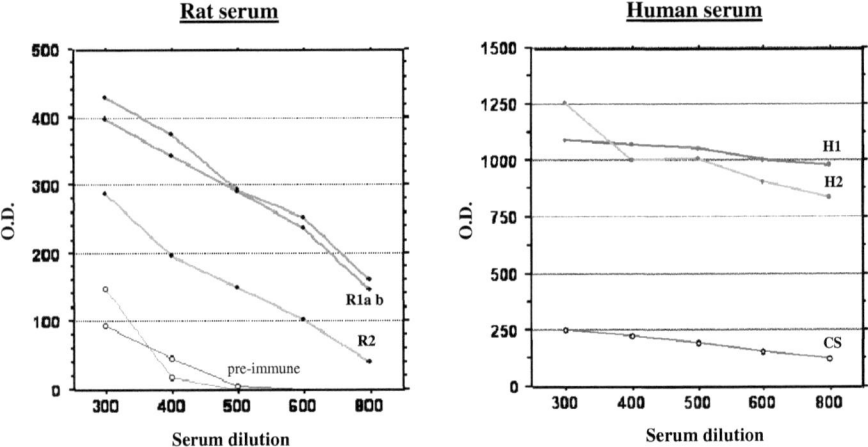

Figure 3.2.9-1 ELISA shows the antibody recognition to iL3 E/S products. Left side shows the sera from two rats (R1, R2) and the lower titration of the pre-immune sera. Right side shows the antibody recognition of two sera from human individuals previously infected with *S. stercoralis* (H1, H2) in relation to a human control serum (CS).

Results 53

3.3 Mass spectrometry

To study the differences in protein expression and secretion of *S. ratti*, samples from different stages were analysed using Liquid Chromatography Tandem Mass Spectrometry (LC-MS/MS). The above mentioned stages iL3, pf and fls were chosen for further investigation. 2-D gel electrophoresis was conducted with iL3 E/S products in preliminary tests (data not shown). But it was shown that the amount of protein needed to produce reproductive gels is too large and might be performed with iL3 E/S products but is hard to be conducted for pf- and fls E/S products. In addition, 2-D gel electrophoresis does not favour the identification of less abundant proteins or those with extreme isoelectric points (pI). Therefore it was decided to perform shotgun LC-MS/MS as a complementary approach, which allows the simultaneous identification of multiple proteins in a complex mixture. The LC-MS/MS analysis was performed at the Proteomics Center at Children's Hospital, Boston, USA. The general workflow for the preparation of samples and the conduction of mass spectrometric analysis is given in figure 3.3-*1*. The parasitic life cycle and all sample preparation steps where performed at the BNI, Hamburg. All gels used for mass spectrometric analysis and the analysis itself were prepared at the Proteomics Center. The databases searches and the data evaluation were performed at the BNI and the Proteomics Center.

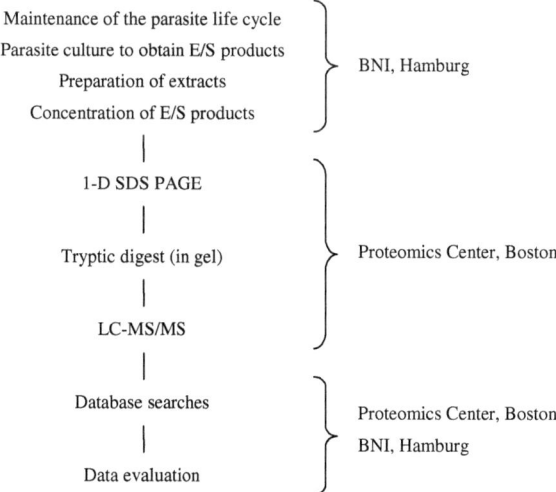

Figure 3.3-*1* General workflow giving an overview of the locations and important steps for performing the mass spectrometric analysis.

3.3.1 Comparison of proteins secreted from different stages

LC-MS/MS was also carried out on whole worm extracts to confirm differences in protein composition compared to E/S products as suggested by 1-D gel electrophoresis. In a preliminary test a "whole worm" analysis was performed to test whether it is possible to detect proteins with relatively low amounts of individuals. For this method larvae were counted and picked individually and subjected to mass spectrometric analysis, as described in 2.3.3.10. 1–1,000 iL3 were counted, 1-D SDS PAGE was performed and the bands were analysed (Figure 3.3.1-*1*).

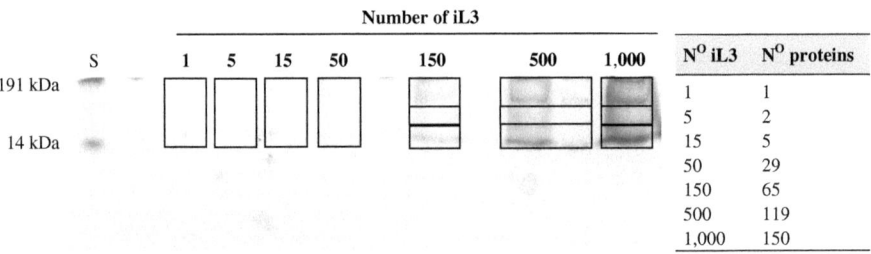

Figure 3.3.1-*1* 1-D SDS PAGE of whole iL3s after heating in loading buffer. Seven different amounts of larvae, shown by the numbers above the gel, have been counted and loaded onto the gel. The boxes indicate the gel pieces that have been cut. The table on the right side shows the number of proteins that have been identified in the respective batch.

Looking at the number of identified proteins in the respective batches a clear increase can be seen. The maximum number of 150 proteins was identified in the last batch with 1,000 larvae, representing a volume of approximately 0.011 mm^3. Due to the effort it takes to prepare the worms and considering that cellular components, membrane proteins and proteins of low abundance might be underrepresented using this approach, it was decided to use the regular native worm extracts for the comparative analysis with E/S products.

By applying LC-MS/MS for the E/S products and extracts of iL3, pf and fls it was possible to identify 1,081 proteins in total. The Venn diagram shows the distribution of identified proteins in the E/S products (Fig 3.3.1-2) and the extracts (Fig 3.3.1-3) within the three stages. For the E/S products 587 proteins have been found in all the stages. 195 proteins, representing roughly one third of the total protein count of all stages, have been found solely in the iL3. 78 proteins (13 %) have been found in the parasitic females and 34 proteins (5 %) have been found in the free-living stages only, whereas 140 proteins (23 %) have been found in samples from every stage.

Results

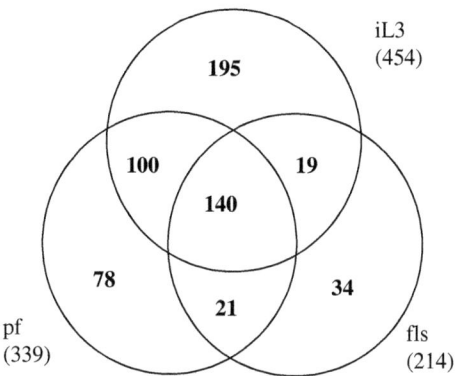

Figure 3.3.1-2 Schematic representation of proteomic data. The Venn diagram shows the distribution of identified *S. ratti* E/S products over three life cycle stages – iL3, pf, and fls. Numbers in brackets show the numbers of identified proteins in each stage in total.

995 proteins have been identified in the extracts of iL3, pf and fls. Here the fls represent the largest number with 723 proteins (73 %) followed by the pf with 574 proteins (58 %) and the iL3 with 489 proteins (49 %). 282 proteins (28 %) have been found only in the fls and almost the same amount of 271 proteins (27 %) in all stages. The lowest number of proteins could be allocated in the iL3 only. Here 68 proteins (9 %) were found and 125 proteins (13 %) were found in the pf only.

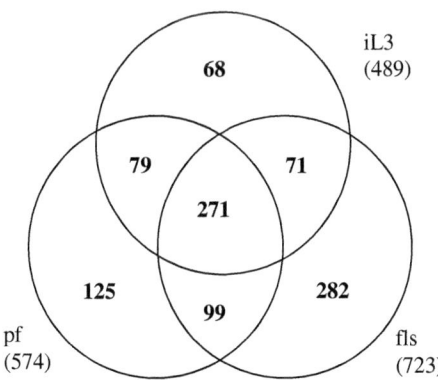

Figure 3.3.1-3 Schematic representation of proteomic data. The Venn diagram shows the distribution of identified *S. ratti* extract proteins of the three life cycle stages – iL3, pf, and fls. Numbers in brackets show the numbers of identified proteins in each stage in total.

The graph in figure 3.3.1-4 shows in percent the composition of the identified proteins in the E/S products and in the extracts for the various stages. By direct comparison of the results found for the E/S products and the extracts it can be seen that there is a large shift from the iL3 to the fls. The fls have a low number of secreted proteins compared to the proteins present in the extract. Whereas the iL3 have a high number of secreted proteins as in the combined iL3 and pf extract proteins. The number of identified proteins in all stages shows only a small increase in the extract whereas the number of proteins found in iL3 and pf is smaller.

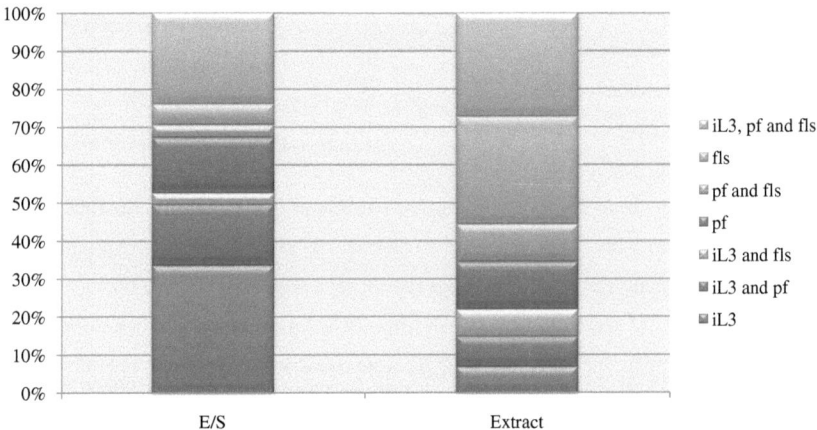

Figure 3.3.1-4 Schematic representation of identified proteins within the different stages. The graph shows in percent the composition of identified proteins in the E/S products and the extracts. The different colours show the single stage or combination of stages in which proteins were found.

The graph in figure 3.3.1-5 shows the compilation of unique and common proteins in the E/S products and extracts. It should be stated that the stage with most proteins being found only in the E/S products is the iL3 followed by the pf and the fls. The fls, however, is the stage with most proteins found in the extracts, followed by the pf and the iL3.

Results 57

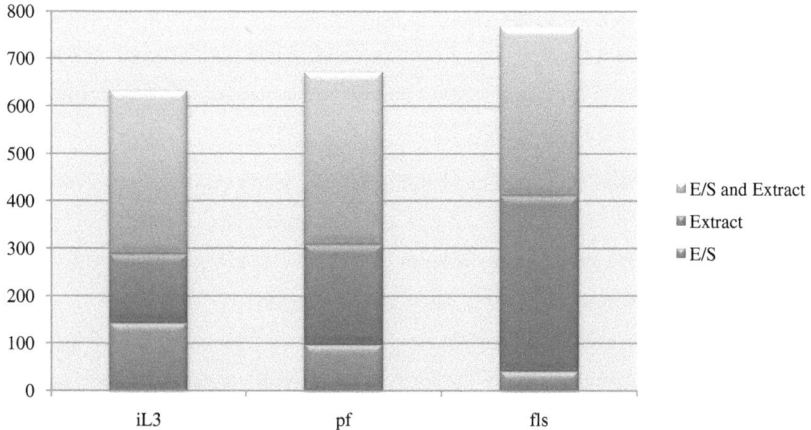

Figure 3.3.1-5 Schematic representation of identified proteins within E/S- and extract proteins. The graph shows the compilation of unique and common proteins in the E/S products and extracts. Bars represent the number of proteins in the different stages.

3.3.2 Abundant proteins in E/S products shared by all stages

Each sample type was analysed in triplicate. In total, roughly 600 mass spectrometric runs were completed. Raw data files were acquired automatically and had to be converted to mascot generic format (mgf) files using a custom file converter. Mgf-data format was required for performing the database searches using the ProteinPilot™ search engine as described in 2.3.4.1. For the data evaluation all proteins having two or more peptide matches were considered as positive hits. Every hit was checked individually for the number of matching peptides to exclude redundancies. For comparison of the respective stages and search results the data then was combined in one excel sheet. To allocate the search data including the EST-numbers to possible protein functions BLAST searches were performed using the protein BLAST program on the NCBI web page. BLAST results were tested for possible signal peptide sequences using SignalP. The data and search results were listed in tables that can be found in annex 9.1 (protein lists).

Table 3.3.2-*1* lists the 25 most abundant E/S proteins that were common in all the stages. The *unused protein score* is a measurement of all the peptide evidence for a protein that is not better explained by a higher ranking protein. It is based on the number of peptides pointing to a certain protein, whereas peptide identifications can only contribute to the UPS of a protein to the extent that their spectra have not already been used to justify more confident proteins. Thus it is the true indicator of protein confidence. In addition, the *unused protein score* gives an estimate of the abundance of a protein. For all identifications in this work proteins with an UPS below 4.00 where excluded. The table consists of ten columns containing the following information:

1. Cluster: Shows the cluster number of a *S. ratti* (SR0xxxx) or a *S. stercoralis* (SS0xxxxx) EST.
2. BLAST alignment: Shows the result of a BLAST search performed with the protein sequence represented by the respective cluster number.
3. Species: Gives the species of the resulting sequence from the BLAST search.
4. Accession number: Shows the corresponding NCBI protein accession number for the BLAST search sequence.
5. E value (E): Shows the resulting expectation value from the BLAST search. The expectation value describes the number of different alignments with scores equivalent to or better than the raw alignment score that are expected to occur in a database search by chance. The lower the E value, the more significant the score.
6. Signal peptide (SP): Shows if the sequence represented by the accession number carries a signal peptide.
7. EST length (EST Lgt.): Shows the sequence length of the EST cluster.
8. Percent coverage (% Cov.): Gives the sequence covered by the identified peptides in percent.
9. Number of peptides (# Pep.): Shows the number of peptides found within a certain EST cluster sequence.
10. *Unused protein score* (UPS): Shows the score based on the number of peptides pointing to an EST cluster sequence.

Some of the proteins listed in table 3.3.2-*1* as well as proteins listed in the supplementary tables were also found in the samples that were recovered from supernatants supplemented with CHX. Thus these proteins might represent proteins that are present through shedding or leakage of dead worms.

Two proteins are assembled within the 25 highest scoring proteins which belong to the heat-shock protein (HSP) family. One of them is the highest scoring protein having a UPS of 71.01. The Cluster number is SR00728 and the BLAST search leads to a HSP-60 sequence from

S. ratti that was submitted to the NCBI protein database in 2007. The second protein, SR01060, relates to HSP-70 having a score of 21.39. Other HSP homologous EST cluster sequences, like SR00952 (HSP 90), SS01752 (HSP 10) and SR00065 (HSP-70c) were found in the E/S products. Besides HSP-60 another protein relating to a *Strongyloides* species was identified. The Cluster SR01871 is with a UPS of 46.28 an abundant protein and showed a BLAST similarity to a protein named L3NieAg used as immunodiagnostic antigen for the detection of *S. stercoralis* infection (Ramanathan, 2008). Some cluster sequences relating to structural proteins, as SS01554 (actin 1), SS01430 (tropomyosin) and SR00774 (ubiquitin) are listed among the 25 highest scoring proteins as well as functional proteins e.g. SR01037 (protein disulfide isomerase), SR00966 (arginine kinase) and SR00792 (probable citrate synthase). Interestingly the table also contains cluster sequences that give BLAST results for hypothetical proteins, like SR00866 and SR00872 that are possibly representing nematode specific, previously not further characterised proteins. Another group of proteins that should also be mentioned are the galectins. These are the cluster numbers SS00840, SR00627 and SR00900 along with SR00857, which is not listed in the 25 highest scoring proteins but has also been found in all stages. They all show similarities to galectins from other parasitic nematodes and also to vertebrates. The results of further analyses will be discussed in section 3.4 of the presented work. In total 140 proteins were identified in E/S samples from all stages. 19 proteins could not be assigned to any function or cellular compartment and four proteins have an E-value of more than $1e^{-10}$. Six proteins do not represent *S. ratti* or *S. stercoralis* cluster numbers but rather proteins from other nematode species, mainly *C. elegans*.

Results

Table 3.3.2-1 The 25 highest scoring proteins found in all stages according to the UPS. Abbreviations are explained in the text. * - proteins were also found in CHX supplemented supernatants, ** - sequence length refers to the complete sequence investigated at the BNI, trun - truncated.

	Cluster	BLAST Alignment	Species	Accession Number	E	SP	EST Lgt.	% Cov.	# Pep.	UPS
1	SR00728	Heat-shock protein 60	S. ratti	ABY65231	0.0	no	563**	64.1	33	71.01
2	SR01037*	Protein disulfide isomerase	A. suum	CAK18511	0.0	yes	499	62.7	27	62.56
3	SR01871*	L3NieAg.01	S. stercoralis	AAD46493	$4e^{-21}$	trun	169	71.0	18	46.28
4	SR00866	Hypothetical protein CBG18957	C. briggsae	XP_001674360	$7e^{-139}$	yes	338	51.5	21	46.19
5	SR00966	Arginine kinase	H. glycines	AAO49799	$2e^{-165}$	no	346	61.8	19	41.48
6	SR01042	Elongation factor 2	C. elegans	P29691	0.0	no	531	29.2	12	33.34
7	SS00840	Galectin 1	T. circumcincta	AAD39095	$4e^{-127}$	no	278	55.8	14	32.70
8	SR00922*	14-3-3 family member	C. elegans	NP_509939	$4e^{-110}$	no	250	59.2	12	32.14
9	SR00876	Calponin protein 3	C. elegans	O01542	$3e^{-55}$	no	144	76.4	13	31.21
10	SR00881	W07G4.4 Peptidase M17	C. elegans	NP_506260	$1e^{-111}$	no	356	31.5	11	25.77
11	SR00627	Galectin	H. contortus	AAF63405	$7e^{-139}$	no	276	38.0	10	25.56
12	SR00858*	Fatty acid binding protein	A. suum	P55776	$7e^{-60}$	yes	165	60.0	12	25.00
13	SS01554*	Actin 1	B. malayi	XP_001895795	$3e^{-55}$	no	376	38.6	11	24.80
14	SR00792	Probable citrate synthase	C. elegans	P34575	$1e^{-166}$	no	339	46.0	11	24.50
15	SS01430*	Tropomyosin	T. pseudospiralis	Q8WR63	$4e^{-124}$	no	284	34.9	10	24.18
16	SR01407	Protein disulfide isomerase	T. circumcincta	Q2HZY3	0.0	yes	287	35.9	10	23.95
17	SR00027	Elongation factor 1-alpha	B. malayi	XP_001896880	0.0	no	455	29.5	9	22.79
18	SR04614	Probable aspartate aminotransferase	C. elegans	Q22067	$5e^{-81}$	no	187	63.6	10	22.78
19	SR00872	Hpothetical protein	C. briggsae	XP_001676876	$1e^{-134}$	no	307	38.8	9	22.47
20	SR00949	ATP synthase subunit	C. elegans	NP_498111	0.0	no	473	34.5	10	21.70
21	SR01060	Heat-shock protein 70	P. trichosuri	AAF87583	0.0	no	644	25.8	10	21.39
22	SR00774*	Ubiquitin	C. elegans	NP_741157	0.0	no	439	84.1	6	18.68
23	SR02163	Peptidase family M1 containing protein	B. malayi	XP_001897028	$4e^{-36}$	no	178	51.7	8	18.51
24	SR00900	Galectin	B. malayi	XP_001896448	$2e^{-68}$	no	163	54.6	6	18.50
25	SR00941	Aconitase family member	C. elegans	NP_498738	$1e^{-123}$	no	280	42.1	8	18.18

3.3.3 Stage-related proteins

3.3.3.1 Proteins enriched in infective larvae

Table 3.3.3.1-*1* lists the 25 highest scoring proteins found only in iL3 E/S products. It was observed that the highest UPS scores for this group of proteins were lower compared to proteins found in all stages. Remarkably the previously mentioned astacin metalloproteinase was identified in none of the pf or fls E/S products, which is consistent with the observation that only the E/S products showed a strong proteolytic activity. In the search database the protease was newly assigned as SR11111 and represents the full length sequence that has first been described in the BNI (Borchert, 2007). Here the BLAST search gave a positive result for a *S. stercoralis* astacin which was previously characterised and found in iL3 cDNA libraries of *S. stercoralis* (Gomez Gallego, 2005). Another EST cluster, SS01564, is the homologous *S. stercoralis* astacin sequence. As it represents only a partial SR11111 sequence its' UPS had lower scores and is not listed in this table. As previously shown it is secreted by the iL3 which is supported by the fact that the sequence carries a signal peptide. As reported in the proteins contained in samples from all stages the iL3 specific proteins also comprise an EST cluster (SR00386) that is related to the immunodiagnostic antigen L3NieAg from *S. stercoralis*. However, the BLAST result is not reliable considering the high E-value of 0.15. Thus this protein might either be a new *Strongyloides* spp. protein or a *S. ratti* specific protein. Some cluster sequences relating to myosin proteins as SR01001, SS01266 and SR00998 are listed among the 25 highest scoring proteins as well as functional proteins e.g. SR01803 (thiosulfate sulfuryltransferase), SR03119 (short chain reductase/dehydrogenase) and SR00383 (propionyl coenzyme A carboxylase). Beside SR00386 two further proteins show high E-values – SR02741 and SR00366. One of them, SR00366, belongs to the group of proteins giving BLAST results for hypothetical proteins together with SS01256 and SR01321. In total 37 out of 195 proteins belong to the group that could not be assigned to any function or cellular compartment with five of them having an E-value of above $1e^{-10}$. 26 proteins do not represent *S. ratti* or *S. stercoralis* cluster numbers but rather proteins from other nematode species, mainly *C. elegans*.

Table 3.3.3.1-1 The 25 highest scoring proteins of iL3 E/S products according to the UPS. Abbreviations are explained in the text.

	Cluster	BLAST Alignment	Species	Accession Number	E	SP	EST Lgt.	% Cov.	# Pep.	UPS
1	SS01511	RAS-related protein RAB-1A	B. malayi	XP_001901944	$7e^{-104}$	no	205	57.5	8	16.05
2	SR02886	Hypothetical 35.6 kDa protein	B. malayi	XP_001899587	$1e^{-71}$	no	180	55.0	7	14.10
3	SR01803	Thiosulfate sulfuryltransferase	B. malayi	XP_001901653	$1e^{-15}$	no	174	53.4	7	14.03
4	SS00138	Adenylate kinase	B. malayi	XP_001894222	$4e^{-62}$	no	149	49.6	6	14.00
5	SR11111	Metalloproteinase precursor	S. stercoralis	AAK55800	$2e^{-61}$	yes	265	43.8	6	13.72
6	SR01001	Myosin – filarial antigen	B. malayi	AAB35044	0.0	trun	490	41.2	13	13.67
7	SR02558	Lethal family member (let-805)	C. elegans	NP_001022641	$1e^{-69}$	yes	190	37.8	5	13.52
8	SR00386	L3NieAg.01	S. stercoralis	AAD46493	0.15	trun	112	50.8	5	13.50
9	SR00366	Hypothetical protein DDBD-RAFT_0217849	D. discoideum AX4	XP_642992	$5e^{-08}$	no	190	40.5	6	13.15
10	SR00901	TPR domain containing protein	B. malayi	XP_001902724	$5e^{-48}$	no	226	34.9	6	12.73
11	SS02590	Sensory Axon guidance family member	C. elegans	NP_001033397	$3e^{-42}$	yes	169	59.7	6	12.00
12	SR03037	Trans thyretin related family member	C. elegans	NP_499054	$1e^{-30}$	yes	147	50.3	5	11.80
13	SR03119	Short chain reductase/dehydrogenase	B. malayi	XP_001900343	$1e^{-46}$	no	179	30.1	5	11.57
14	SS01266	Myosin-4	C. elegans	P02566	$3e^{-94}$	no	258	26.0	6	11.45
15	SR00998	Myosin light chain family member	C. elegans	NP_510828	$2e^{-73}$	no	170	31.8	4	10.60
16	SR04474	Peptidase family M1 containing protein	B. malayi	XP_001897028	$2e^{-64}$	no	196	32.1	4	10.36
17	SR03753	K02D10.1b	C. elegans	NP_498936	$5e^{-30}$	no	154	14.3	3	10.05
18	SR02741	Fatty acid retinoid binding protein	W. bancrofti	AAL33794	0.37	no	139	15.8	4	10.02
19	SS01256	Hypothetical protein Bm1_13900	B. malayi	XP_001894244	$1e^{-26}$	yes	228	39.0	5	10.00
20	SR01321	Hypothetical protein Bm1_36850	B. malayi	XP_001898817	$1e^{-12}$	no	176	43.2	5	10.00
21	SR00700	Na, K-ATPase alpha subunit	C. elegans	AAB02615	$3e^{-101}$	no	233	29.2	4	9.71
22	SR02060	Cell division cycle related family member	C. elegans	NP_495705	$3e^{-92}$	no	193	32.6	4	9.50
23	SR03954	CAP protein	B. malayi	XP_001891888	$2e^{-53}$	no	242	28.1	4	9.22
24	SS01276	F55F3.3	C. elegans	NP_510300	$2e^{-104}$	no	317	12.0	3	9.17
25	SR00383	Propionyl Coenzyme A Carboxylase	C. elegans	NP_509293	$2e^{-17}$	no	76	50.0	3	9.10

3.3.3.2 Proteins enriched in parasitic females

The 25 highest scoring proteins found only in parasitic female E/S products are listed in table 3.3.3.2-1. The highest scoring protein is represented by the EST cluster SR01608 which relates to an EF-hand protein family member from *B. malayi* according to the BLAST result and carrying a signal peptide according to SignalP. With eight detected peptides and a coverage of 69.0% for the 158 amino acid fragment is an abundant protein found in the parasitic females. It was, however, also found in iL3 extracts with a lower UPS. Further EST clusters relating to *B. malayi* proteins are SR00396, SR02118, SS03344, SR01499 and SR00986. Two of these proteins are also carrying a signal peptide. SS03344 was blasted as SPARC precursor with an E-value of $2e^{-74}$. In *C. elegans* the SPARC protein has been shown to be an extracellular calcium-binding protein (Schwarzbauer, 1994). The second cluster SR00396 was blasted as endoplasmin precursor with an E-value of $7e^{-98}$. Endoplasmins belong to the group of HSPs and function in the processing and transport of secreted proteins. The human endoplasmin precursor for example is also termed HSP90 beta. Two other cluster numbers have been related to HSPs. SR00984 gave a BLAST result for a small HSP from the parasitic nematode *T. spiralis* (E-value $2e^{-21}$) that has been shown to be present in *T. pseudospiralis* E/S products (Ko, 1996), and the BLAST search for SR03349 resulted in HSP-17 from *C. elegans* (E-value $2e^{-20}$). Sequence alignment showed a homology of 49 % indicating that the sequences might represent a HSP family with relevance for parasitism within *S. ratti*. Out of the two proteins classified as hypothetical proteins one had a high E-value of 2.0 (SR04455). Besides that two further proteins showed E-values >$1e^{-10}$, one of them, cluster number SR04713, was blasted as BspA-like surface antigen from *T. vaginalis* (E-value 5.3) and the second cluster, SR03587, was blasted as a metalloprotease from *N. vitripennis*. SR02663 and SR03310 are further sequences showing similarities to metalloproteinase precursors. Although the *N. vitripennis* metalloprotease is specified as an astacin like metalloprotease, the similarities to SR03587, SR02663, SR03310 and the previously mentioned SR11111 were only between 16 and 19 %. Other protease like sequences were SR03901 (aspartyl protease), SR03191 and SR01641 both showing similarities to POPs. The sequences showing similarities to POPs were subject to further studies presented in chapter 3.4.2 of this work. In total 78 proteins were identified solely in the parasitic female E/S products of which 11 were not assigned to any function or specific protein and eight proteins did not relate to a *S. ratti* or *S. stercoralis* EST but to other sequences from *C. elegans*, *N. americanus* and *H. glycines*. The characterisation of those proteins could result in an identification of new functional secreted proteins.

Table 3.3.3.2-1 The 25 highest scoring proteins of pf E/S products according to the UPS. Abbreviations are explained in the text.

	Cluster	BLAST Alignment	Species	Accession Number	E	SP	EST Lgt.	% Cov.	# Pep.	UPS
1	SR01608	EF hand family protein	B. malayi	XP_00190111	$2e^{-37}$	yes	158	69.0	8	23.28
2	SR03191	Prolyl endopeptidase	R. norvegicus	EDL99674	$3e^{-19}$	no	189	51.3	6	16.69
3	SR03587	Metalloproteinase	N. vitripennis	XP_001606489	$9e^{-06}$	yes	166	42.8	6	14.23
4	SR03901	Aspartyl protease (asp-2)	C. elegans	NP_505384	$4e^{-43}$	no	191	56.5	5	13.20
5	SR04847	Acetylcholinesterase 2	D. destructor	ABQ58116	$1e^{-44}$	no	192	36.5	5	12.4
6	SR03310	mp1	O. volvulus	AAV71552	$2e^{-13}$	yes	189	39.2	5	11.64
7	SR01641	Prolyl endopeptidase	T. denticola	NP_971802	$2e^{-12}$	no	126	31.0	4	11.26
8	SR00564	Calumenin	C. elegans	NP_001024806	$2e^{-134}$	yes	286	19.2	4	10.87
9	SR00984	Small heat-shock protein	T. spiralis	ABJ55914	$2e^{-21}$	no	160	35.6	4	10.67
10	SR04455	Hypothetical protein CBG05204	C. briggsae	XP_001664881	2.0	yes	88	56.8	5	10.64
11	SR02054	Scavenger receptor cysteine rich protein	C. pipiens quinquefasciatus	XP_001866937	$3e^{-24}$	yes	328	14.6	4	10.44
12	SR03349	Heat-shock protein HSP17	C. elegans	NP_001023958	$2e^{-20}$	no	157	39.5	5	10.04
13	SR01073	Ribosomal protein (rpl-5)	C. elegans	NP_495811	$2e^{-119}$	no	290	19.3	4	9.47
14	SR00396	Endoplasmin precursor	B. malayi	XP_001899398	$7e^{-98}$	yes	231	16.5	3	8.56
15	SR00986	60S ribosomal protein L10	B. malayi	XP_001898297	$2e^{-85}$	no	189	17.5	3	8.26
16	SR01297	Immnunosuppressive ovarian message protein	A. suum	CAK18209	$2e^{-17}$	yes	324	19.8	4	8.15
17	SR04713	Surface antigen BspA-like	T. vaginalis	XP_001315000	5.3	no	55	89.1	3	7.27
18	SR02663	Metalloproteinase precursor	S. stercoralis	AAK55800	$6e^{-12}$	trun	182	19.2	3	6.91
19	SR01002	Ribosomal protein (rps-18)	C. elegans	NP_502794	$5e^{-75}$	no	154	15.6	2	6.64
20	SR00979	Ribosomal protein L9	S. papillosus	ABK55147	$2e^{-73}$	trun	166	21.7	2	6.47
21	SR02153	Hypothetical protein CBG20335	C. briggsae	CAP37373	$2e^{-34}$	no	178	13.5	2	6.34
22	SR02118	Phosphoribosyl transferase	B. malayi	XP_001895434	$2e^{-27}$	no	152	31.6	3	6.24
23	SS03220	Intermediate filament protein (ifa-3)	C. elegans	NP_510649	$6e^{-38}$	no	141	24.1	3	6.23
24	SS03344	SPARC precursor	B. malayi	XP_001897784	$2e^{-74}$	yes	175	20.6	3	6.23
25	SR01499	Troponin family protein	B. malayi	XP_001898461	$5e^{-50}$	no	257	14.0	3	6.19

3.3.3.3 Proteins enriched in the free-living stages

Table 3.3.3.3-*1* lists the 25 highest scoring proteins identified in samples from the free-living stages. In total, interestingly, only 33 proteins were found in E/S products from the free-living stages. On the other hand 221 proteins were found in the extract samples prepared from the free-living stages. 11 of the 25 highest scoring proteins showed alignments to hypothetical proteins from *C. elegans*, *C. briggsae* and *B. malayi*. The E-value of 7.9 for the cluster number SR00380 showed an unspecific alignment for a hypothetical protein from *E. ventricosum*. Also the clusters SR01169 and SR00821 were unspecific alignments with an aminotransferase from *C. botulinum* and a saposin-like protein from *C. elegans* and thus raised the number of unclassified proteins to 13. Also the three unspecific clusters may represent proteins that are unique to *Strongyloides* and that are important and specific to one or more non-parasitic stage. The cluster number SS00929 aligned to a high mobility group box (HMGB) protein from *C. elegans*. In both *C. elegans* and *B. malayi* proteins of the HMGB family were shown to be expressed predominantly in developing larvae (Jiang, 2008), thus this protein may be derived from the presence of L1 and L2 larvae in the cultures. The usage of the amino acid substitution mode led to the identification of 22 proteins originally belonging to published *C. elegans* and *C. briggsae* listed in table 12b of section seven.

Table 3.3.3.3-*1* The 25 highest scoring proteins of fls E/S products according to the UPS. Abbreviations are explained in the text.

	Cluster	BLAST Alignment	Species	Accession Number	E	SP	EST Lgt.	% Cov.	# Pep.	UPS
1	SR02994	Hypothetical protein Y49E10.18	*C. elegans*	NP_499623	$2e^{-27}$	yes	143	51.0	7	19.13
2	SR00863	MFP2b	*A. suum*	AAP94889	$5e^{-71}$	no	173	52.0	7	14.85
3	SR00375	Hypothetical protein CBG05949	*C. briggsae*	XP_001670383	$5e^{-09}$	yes	148	31.1	5	13.91
4	SR02091	Hypothetical protein CBG22129	*C. briggsae*	XP_0016676 27	$1e^{-35}$	yes	174	36.8	6	13.27
5	SR00671	Lysozyme family member (lys-5)	*C. elegans*	NP_502193	$4e^{-37}$	yes	160	23.1	4	12.70
6	SR02511	Acyl sphingosine amino hydrolase	*C. briggsae*	CAP33700	$2e^{-48}$	yes	184	27.2	4	8.35
7	SR01169	Aminotransferase	*C. botulinum*	ZP_02614737	0.53	no	176	31.8	4	8.14
8	SR00576	MSP domain protein	*B. malayi*	XP_0018996 79	$3e^{-32}$	no	97	47.4	4	8.02
9	SR00479	Hexosaminidase B	*P. troglodytes*	XP_517705	$2e^{-46}$	yes	167	16.8	2	7.13
10	SR00767	F25A2.1	*C. elegans*	NP_503390	$6e^{-25}$	no	178	17.4	2	6.67
11	SS01173	Enoyl-CoA reductase	*A. suum*	AAC48316	$1e^{-104}$	no	299	10.0	2	6.41

Table 3.3.3.3-1 continued

Cluster	BLAST Alignment	Species	Accession Number	E	SP	EST Lgt.	% Cov.	# Pep.	UPS
12 SR00821	Saposin-like protein	C. elegans	NP_491803	5.4	yes	86	54.7	3	6.39
13 SR00750	Similar to mannose receptor	G. gallus	XP_418617	$2e^{-07}$	yes	174	21.3	3	6.35
14 SR00354	Acid sphingo-myelinase	C. elegans	NP_001040996	$2e^{-89}$	yes	269	19.3	3	6.09
15 SR01936	Hypothetical protein CBG21853	C. briggsae	XP_0016727 42	$1e^{-21}$	yes	190	13.7	2	5.98
16 SR00380	Hypothetical protein EUBVEN_01944	E. ventri-osum	ZP_02026680	7.9	yes	154	14.9	2	5.78
17 SR05257	Putative serine protease F56F10.1	C. elegans	P90893	$2e^{-24}$	yes	185	25.4	2	5.52
18 SR02550	Putative serine protease F56F10.1	C. elegans	P90893	$2e^{-35}$	yes	239	11.7	2	5.22
19 SS01082	Hypothetical 86.9 kDa protein	B. malayi	XP_0018960 95	$5e^{-51}$	no	309	6.8	2	5.22
20 SR02018	Yeast Glc seven-like Phosphatase	C. elegans	NP_491237	$2e^{-94}$	no	184	13.0	2	4.74
21 SR00716	F09C8.1	C. elegans	NP_510636	$3e^{-45}$	yes	170	22.4	2	4.66
22 SR01063	Aspartyl protease precursor	C. briggsae	CAP30637	$2e^{-98}$	yes	359	10.9	2	4.49
23 SS00929	High mobility group protein	C. elegans	NP_496970	$5e^{-21}$	no	94	20.2	2	4.02
24 SR00223	Hypothetical protein C50B6.7	C. elegans	NP_506303	$6e^{-51}$	yes	192	18.2	2	4.02
25 SR00899	Hypothetical protein CBG09313	C. briggsae	XP_0016742 44	$4e^{-11}$	no	231	13.9	2	4.01

3.3.4 Demonstration of differentially expressed proteins applying PCR analysis

To examine whether abundant proteins identified in single stages reflect the same profile on RNA level single proteins were tested by PCR analysis with cDNA obtained from iL3 and parasitic females. PCR was performed with primers from three different genes all representing coding sequences for proteins found in E/S products. Figure 3.3.4-1 shows that S. ratti galectin-5 (Primers: SrGal5f2 and SrGal5r2) was amplificated in cDNAs from both stages, whereas the proteolytic astacin was only present in iL3 but not in parasitic female cDNA. Inversely the POP could only be found in parasitic female cDNA, which is consistent with the results found in the MS analysis of E/S products from iL3 and parasitic females.

Results 67

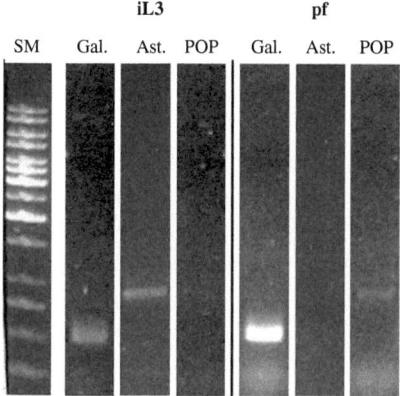

Figure 3.3.4-1 PCR expression analysis of different genes with cDNA from iL3 and parasitic females revealing stage specific expression. 3 proteins were chosen for PCR analysis: (1) Galectin-5 (**Gal.**) was found in E/S products of all stages, (2) Astacin (**Ast.**) was found in iL3 E/S products only, (3) Prolyl-oligopeptidase (**POP**) was found in parasitic female E/S products only. **SM** – size marker.

3.4 Selected candidate functional proteins

During the evaluation process various proteins where chosen for further investigation. The results for two of these proteins or protein groups, respectively, are presented below.

3.4.1 Identification and analysis of *S. ratti* galectins

Since galectins represent an abundant protein family being secreted by *S. ratti* iL3, pf and fls it was decided to further elucidate the sequences and the properties of these molecules. In a first step the *S. ratti* and *S. stercoralis* EST databases where screened for more sequence information of possible nucleotide sequences representing partial or whole galectin structures. For this process an annotated EST database, kindly supplied by Makedonka Mitreva from the Genome Sequencing Center, Washington University School of Medicine, St. Louis, MO, USA, was used. After spotting the EST cluster numbers it was possible to retrace the associated EST nucleotide sequences by using the cluster history and the cluster summary .031211 from *S. ratti* and .011025 from *S. stercoralis* that can be found on the file transfer protocol (FTP) server on nematode.net (http://www.nematode.net/FTP/cluster_ftp/index.php).

In total 12 EST clusters showing galectin homologies were found, six for *S. ratti* and six for *S. stercoralis* (Tab 3.4.1-1). The EST nucleotide sequences were retrieved at the NCBI nucleotide database using the according accession numbers. Where more than one EST was available the nucleotides were aligned and the longest coherent sequences were determined.

For the *S. stercoralis* cluster SS00840 and SS01190 primers were designed, used for PCR experiments with *S. ratti* iL3 cDNA and led to PCR products that have been sequenced. The resulting sequences were termed SRX0840 and SRX1190, leading to eight galectin sequences for *S. ratti* in total. The sequences from *S. ratti* were compared to *S. stercoralis* sequences by multiple sequence alignment and it was shown that the sequences match pairwise with regard to the homology (Tab 3.4.1-2). For SR05051 it was found that it was a partial sequence from the previously determined sequence SRX0840 (as shown in brackets in table 3.4.1-2). A nomenclature was proposed where the numbers at the end of the names were chosen because of the highest similarity to *C. elegans* galectins Lec1–Lec15 by direct sequence alignment. However the sequences of SR00838 and SRX1190 could not clearly be assigned to a corresponding *C. elegans* galectin and where thus named *Sr*-Gal-21 and -22. By sequence comparison and analysis of the EST translations it was determined if any 3'- or 5'-ends were present or missing.

Table 3.4.1-1 *S. ratti* and *S. stercoralis* EST clusters and the according nucleotide accession numbers showing homologies with galectis.

		EST Cluster Number	EST Nucleotide Accession Number		
S. ratti	1.	SR00257	kt88a05.y1	kt28e10.y3	
	2.	SR00627	kt18b07.y1	kt51e01.y4	ku22g05.y1
	3.	SR00838	kt34h02.y1 kt37h03.y1	kt92d01.y1 kt37g08.y1	kt30e02.y1
	4.	SR00857	kt47a10.y3 kt57a12.y3	kt21d03.y1 kt62c04.y1	kt13g12.y2
	5.	SR00900	kt12a11.y2 kt83e02.y1	kt20h11.y1 kt17e09.y1	kt83d03.y1 kt47a03.y3
	6.	SR03632	ku43e08.y1		
	7.	SR05051	kw21c04.y1		
S. stercoralis	1.	SS00593	kq13e08.y1	kq19f08.y1	
	2.	SS00732	kq06g10.y1	kq05g05.y1	kq14d03.y1
	3.	SS00840	kq56h03.y1	kp80c05.y2	kp55f06.y1
	4.	SS01127	kq16c01.y1 kq63c01.y1	kq25c07.y1	kq07d05.y1
	5.	SS01190	kp70b05.y1 kp65b07.y1	kp67a11.y1 kp84e10.y2	kp50f11.y1
	6.	SS02581	kp46d10.y1		
	7.	SS02629	kq17g03.y1		

Results

Table 3.4.1-2 Pairwise comparison of *S. ratti* and *S. stercoralis* galectin sequences. The table also includes the proposed nomenclature of the *S. ratti* galectins and shows the presence or absence (+/-) of 3' or 5'-ends. The homology represents the alignment score.

Assigned Name	*S. ratti* EST Cluster	3'	5'	*S. stercoralis* EST Cluster	3'	5'	Homology
Sr-Gal-1	SR00627	-	+	SS00732	-	-	96
Sr-Gal-2	SRX0840 (SR05051)	+ -	+ +	SS00840	+	+	96 (93)
Sr-Gal-3	SR00900	-	-	SS00593	-	-	90
Sr-Gal-5	SR00857	-	-	SS02629	-	-	83
Sr-Gal-11	SR00257	+	-	SS01127	-	-	94
Sr-Gal-21	SR00838	-	?	/	/	/	/
Sr-Gal-22	SRX1190	+	+	SS01190	+	+	90

Since seven sequences overall could be allocated to the galectin family and only four sequences were found in E/S products and extracts, PCR analysis was performed with all galectins using cDNA from iL3 (Figure 3.4.1-1). Primers were designed for all galectins covering a sequence of about 200 base pairs. Since the PCR was performed with reverse transcribed mRNA the results show that all galectins are expressed in iL3.

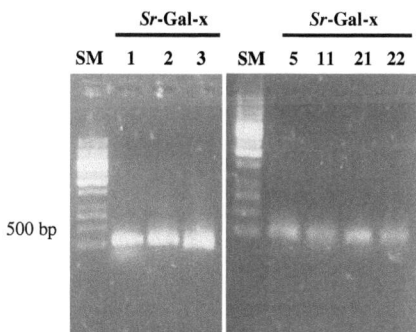

Figure 3.4.1-1 PCR analysis of all galectins found in the *S. ratti* EST database including the newly found *Sr*-Gal-22. As shown every galectin sequence was amplified in iL3 cDNA. **SM** – size marker

3.4.1.1 Completion of galectin sequences

Since only the sequences SS00840 and SS01190 already resembled the full length sequences in the EST database containing a start and stop codon and others were missing one or both ends the aim was to complete at least one or more galectin sequences from *S. ratti*. First efforts using RACE-PCR technique were not successful. Therefore another slightly modified approach was chosen for the completion of the 5'-end. Additional 3'-ends were captured using the ubiquitous nematode spliced leader sequence SL-1 (Figure 3.4.1.1-*1*). The RNA was prepared from 100,000 freshly harvested iL3 and the reverse transcription was performed using the T7I oligo-dT-primer. The resulting cDNA was used for PCR experiments using the SL-1 forward primer and gene specific primers for all seven galectin sequences as reverse primers. The PCR products were cloned into the pGEM-T Easy vector and sequenced. Similarly two further 3'-ends were sequenced (*Sr*-Gal-1 and *Sr*-Gal-3) besides the previously known 3'-end for *Sr*-Gal-2 (SRX0840).

Additional 5'-ends were captured using the T7II reverse primer that is partially complementary to the T7I primer attached to the iL3 cDNA. The respective gene specific galectin forward primers SrGalxfx primers were used as forward primers. Again the positive PCR products were cloned into the pGEM-T Easy vector and sequenced. Equally, two further 5'-ends were identified.

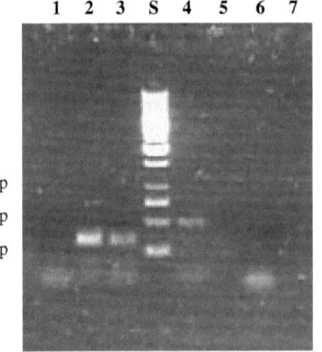

Figure 3.4.1.1-*1* Galectins carrying the SL-1 sequence. The SL-1 primer was used as forward primer for all samples and SrGalrx primers were used as gene specific reverse primers.
Positive: **2** – *Sr*-Gal-1, **3** – *Sr*-Gal-3 and **4** – *Sr*-Gal-2
Negative: **1** – *Sr*-Gal-5, **5** – *Sr*-Gal-21, **6** – *Sr*-Gal-11 and **7** – *Sr*-Gal-21

Results 71

Overall full length or partial sequences have been identified for the following galectins:
- Found in E/S products of all stages: *Sr*-Gal-1 – full length
- Found in E/S products of all stages: *Sr*-Gal-2 – full length
- Found in E/S products of all stages: *Sr*-Gal-3 – full length
- Found in E/S products of all stages: *Sr*-Gal-5 – partial, 3'-end missing
- Not found in E/S products: *Sr*-Gal-11 – partial, 5'-end missing
- Not found in E/S products: *Sr*-Gal-22 – full length

No results were obtained for *Sr*-Gal-21 which has not been identified in *S. ratti* E/S products. For the corresponding cluster number the 3'-end did not exist and by sequence comparison it could not be determined whether the 5'-end might represent the actual start codon.

3.4.1.2 Sequence analysis of galectins

The full length sequences of the secreted galectins *Sr*-Gal-1, -2 and -3, the partial sequence of the secreted Galectin *Sr*-Gal-5 and the full length sequence of *Sr*-Gal-22 were now further analysed. The analogous nucleotide sequences including the translated protein sequences can be

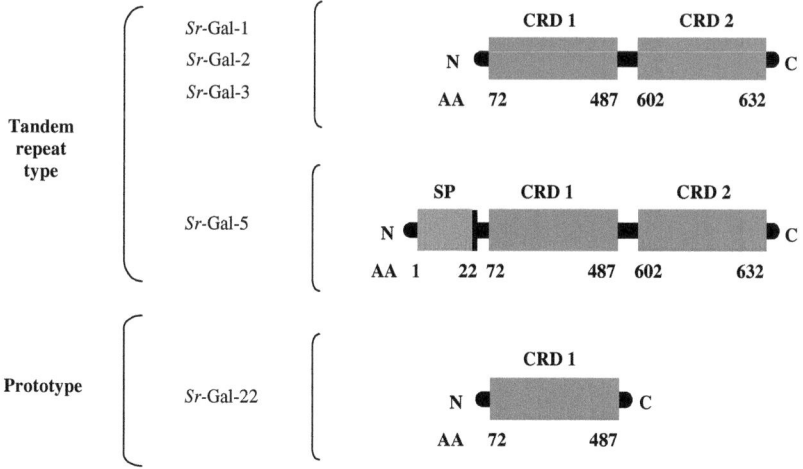

Figure 3.4.1.2-1 Scheme of the domain structure of the *S. ratti* galectins. **AA** – amino acid, **SP** – signal peptide, **CRD** – carbohydrate recognition domain

found in appendices 9.2.1–9.2.4. The domain structure of the galectins was analysed aided by tools for the prediction of motifs using the *Database of Protein Domains, Families and Functional Sites* (http://us.expasy.org/prosite/) and the *Eukaryotic Linear Motif Resource* (http://elm.eu.org/links.html) as well as a signal peptide prediction tool (http://www.cbs.dtu.dk/services/SignalP/).

The domain structures in figure 3.4.1.2-*1* show that *Sr*-Gal-1, -2, -3 and -5 are all tandem repeat type galectins. *Sr*-Gal-5 differs from the other galectins by carrying a signal sequence as predicted by SignalP. *Sr*-Gal-22 is a prototype galectin having only one CRD whereas the tandem repeat type galectins are consisting of two homologous CRDs connected by a linker peptide. The CRD domains are between 120 and 135 amino acids long as shown in the alignment of sequences *Sr*-Gal-1, -2, -3 and *Sr*-Gal-5 (Figure 3.4.1.2-2). The alignment also shows that *Sr*-Gal-5 is carrying a putative signal sequence at the N-terminal end and that it does not have the same degree of homology as the first three sequences. The homologies of the first three sequences range between 60% and 70% whereas the homologies of *Sr*-Gal-5 compared to *Sr*-Gal-1, -2 and -3 are between 38% and 41%. The sequences were used to predict the tertiary structure on the Swiss-Model server (http://swissmodel.expasy.org/). All of the single CRDs showed the

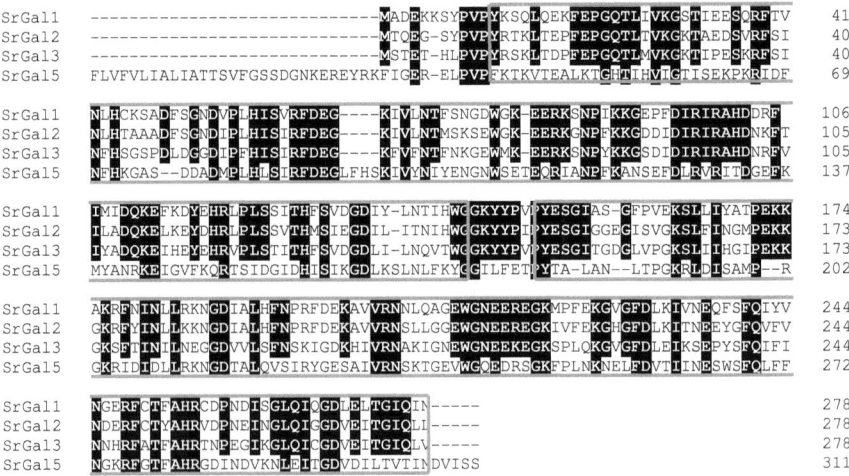

Figure 3.4.1.2-2 Multiple sequence alignment of *Sr*-Gal-1, -2, -3 and -5. Shaded background shows homologous amino acids and the green boxes show the CRDs 1 and 2.

characteristic beta-sandwich structure where the two sheets are slightly bent with six strands forming the concave side and five strands forming the convex side. The concave side forms a groove in which the carbohydrate is bound and which, by comparison to the corresponding template CRDs, is long enough to hold about a linear tetrasaccharide. Figure 3.4.1.2-*3* shows the two CRDs of *Sr*-Gal-3 each from two different views where the bent beta sheets and the groove can easily be seen.

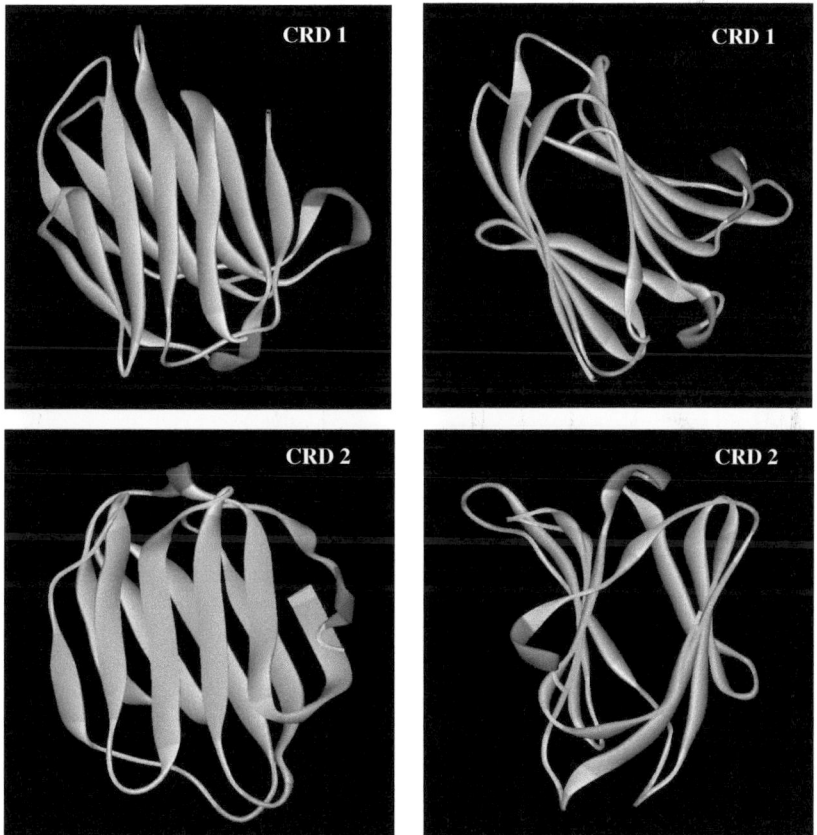

Figure 3.4.1.2-*3* Tertiary structure prediction of the two CRD regions from *Sr*-Gal-3. Light blue ribbons show the bent beta sheets.

3.4.1.3 Phylogenetic analysis of galectins

To address the question if the members of the *S. ratti* galectin family have orthologues in other nematode species a phylogenetic analysis was performed as described in section 2.3.4.2. The search in the NCBI protein database revealed several galectins from other nematode species. *C. elegans* galectins *Ce*-Lec-1–*Ce*-Lec-11 were used as representatives for a non-parasitic nematode and 15 sequences from eight different parasitic nematode species were used in addition. The

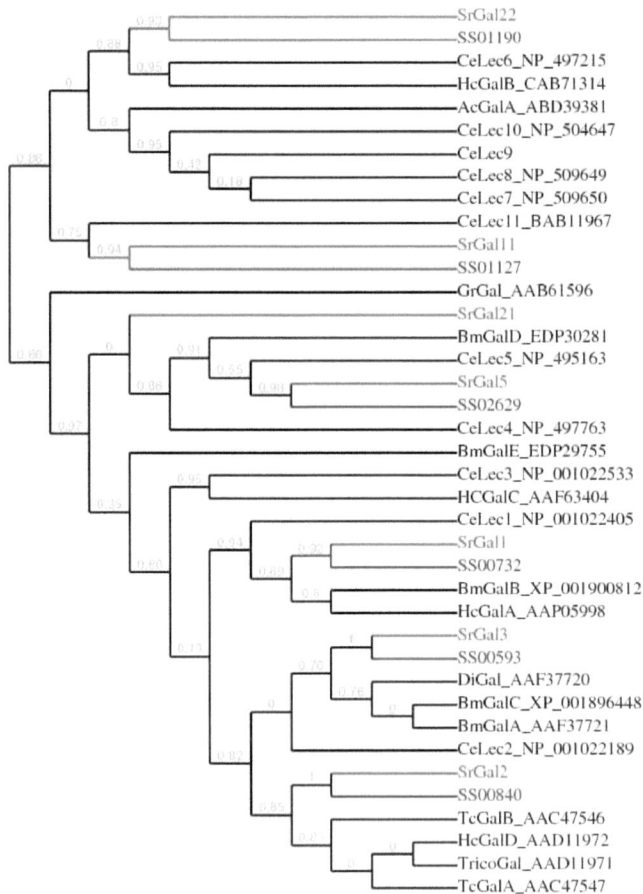

Figure 3.4.1.3-1 Phylogenetic tree of nematode galectins. **Ac** – *Ancylostoma ceylanicum*, **Bm** – *Brugia malayi*, **Ce** - *Caenorhabditis elegans*, **Di** – *Dirofilaria immitis*, **Gr** – *Globodera rostochiensis*, **Hc** – *Haemonchus contortus*, **Tc** – *Teladorsagia circumcincta*, **Trico** – *Trichostrongylus columbriformis*

resulting pylogenetic tree is shown in figure 3.4.1.3-*1*. Six of the *S. ratti* sequences matched to one of the *S. stercoralis* EST cluster sequences as shown in table 3.4.1-*2*. Only *Sr*-Gal-21 does not have an *S. stercoralis* homologue. Each of the galectin sequences is located in a separate branch. *Sr*-Gal-1, *Sr*-Gal-2, *Sr*-Gal-3, *Sr*-Gal-5 and *Sr*-Gal-11 are in the same branch with their respective *C. elegans* homologue.

3.4.1.4 Isolation of native galectins

To show the binding properties of native galectins *S. ratti* galectins, lactose affinity separation was performed with extracts as described in 2.3.3.7. In this experiment the extract proteins were incubated with agarose beads that had lactose covalently bound to their surface. Lactose is a carbohydrate and consists of two monosaccharides. The carbohydrates are β-D-galactose and α/β-D-glucose linked by a β-1,4-glycosidic bond (Figure 3.4.1.4-*1*). The structure of galactose being bound via a β-1,4-glycosidic bond to another carbohydrate molecule is necessary for the binding to galectins.

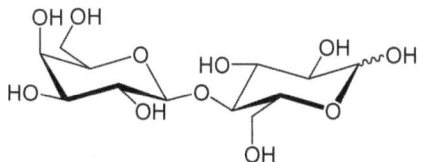

Figure 3.4.1.4-*1* Chair form of lactose with the characteristic β-1,4-glycosidic bond.

After the isolation procedure eluates 1-3 were separated by 1-D-SDS PAGE and stained with coomassie solution. Eluate three was the fraction in which the beads were cooked in SDS loading buffer and the only fraction that showed a slight band detected between 30–35 kDa. Due to the pale visibility of the bands after repeated testing, the gels are not shown. The bands were cut and proteins submitted to tryptic digest and sequencing. The results show that the band contained seven different cluster numbers of which three were galectins (Table 3.4.1.4-*1*). The highest scoring protein was *Sr*-Gal-1 with a UPS of 37.77 followed by *Sr*-Gal-2 with a UPS of 26.31. *Sr*-Gal-3 had a UPS of 7.68. The other proteins were the cluster numbers SR01871, SS01564, SS01499 and SR00990 with different BLAST alignment results. The cluster number SR01871 refers to the *S. stercoralis* immunodiagnostic antigen L3NieAg.01 which was also very frequently found in the E/S proteins from all stages.

The peptide results can be found in appendix 9.2.5. The sequence coverages are shown in figure 3.4.1.4-2.

Table 3.4.1.4-*1* Protein sequences analysed in the bound fraction eluate three.

Result	BLAST alignment / Description	UPS
Sr-Gal-2	Galectin 2	37.77
Sr-Gal-1	Galectin 1	26.31
SR01871	L3NieAg.01 – *S. stercoralis*	12.12
SS01564	Metalloprotease – *S. stercoralis*	9.23
Sr-Gal-3	Galectin-3	7.68
SS01499	Troponin family member – *C. elegans*	6.23
SR00990	Peptidyl-prolyl cis-trans isomerase – *D. immitis*	4.03

Sr-Gal-2: Cov. 56 %

MTQEGSYPVPYR**TKLTEPFEPGQTLTVK**GKTAEDSVRFSINLHTAAADFSGNDIPLHISIR**FDEGKIVLNTMSKSEWG
KEER**KCNPFKKGDDIDIRIR**AHDNKFTILADQK**ELFEYDHR**LPLSSVTHMSIEGDILITNIHWGGKYYPIPYESGIGG
EGISVGKSLFINGMPEKK**GKPR**FYINLLKKNGDIALHFNPR**DEKAVVR**NSLLGGEWGNEEREGK**IVPEKGHGFDLK**IT
NEEYGFQVFVNDER**FCTYAHRVDPNEINGLQIGGDVEITGIQLL

Sr-Gal-1: Cov. 42 %

MADEKKSYPVPYKSQEQER**FEPGQTLIVKGSTIEESQR**FTVNLKCK**SADFSGNDVPLHISVR**TDEGK**IVLNTFSNGDW
GKEER**KSNPIKKGEPFDIRIR**AHDDRFQIMIDQK**LFKDYEHRLPLSSITHFSVDGDIYLNTIHWGGK**YYPVPYESGIA
SGFPVEKSLLIYATPEKK**AKRFMINLLR**KNGDIALHFNPR**DEKAVVR**NNLQAGEWGNEER**EGKMPFEKGVGFDLKIV
NEQFGFQIYVNGERFCTYAHRCDPNDISGLQIQGDLELTGIQIN

Sr-Gal-3: Cov. 17 %

MGTETHLPVPYRSKLTDPFEPGQTLMVFGKTIPESKRFSINFHSGSPDLDGGDIPFHISIRFDEGKFVFNTINKGFWM
KEERKSNPYKKGSDIDIRIRAHDNR**FVIYADQK**EIHEYEHRVFLSTIIHFSVDGDLILNQVTWGGK**YYPVPYESGITG
DGLVPGKSLIIHGIPEKK**GKSFTINILNECGDVVLSFNCKIGDKHIVFNAK**IGNEWGNEEK**EGKSPLQKGVGFDLEIK
SEPYSPQIFINNHRFATFAHRTNFEGIKGLQICGDVEITGIQLV

Figure 3.4.1.4-2 Sequence coverages of *Sr*-Gal-1, -2 and -3 obtained in lactose–agarose bead separation of E/S products. Green peptides have a coverage of 99 and yellow peptides ≥50 - <99. The calculated coverages for each galectin only refer to the green marked peptides.

Results

3.4.1.5 Prokaryotic expression of *Sr*-Gal-3

Sr-Gal-3 was chosen for expression because it was one of the galectins found to be secreted by *S. ratti* and because of the high sequence similarity to the *S. stercoralis* EST cluster sequence. The expression was performed as described in section 2.3.2.4. The expression vector pJC45 contained a histidine coding sequence resulting in a his-tagged protein at the N-terminal end that was purified using Ni-NTA affinity chromatography. Figure 3.4.1.5-*1* shows the different fractions eluted from the affinity column with the *Sr*-Gal-3 band appearing at approximately 33 kDa on the lanes six to eight, whereas fraction eight shows highly purified *Sr*-Gal-3. After dialysis and concentration the corresponding fractions were separated by 1-D SDS PAGE and stained with Coomassie Brilliant Blue solution. The bands were cut, subjected to tryptic digest and sequenced by LC-MS/MS. The result verified the identity of the *Sr*-Gal-3 sequence (data not shown).

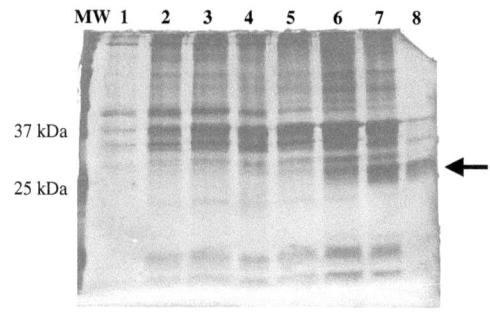

Figure 3.4.1.5-*1* Silver staining of a 1-D SDS PA gel shows the protein fractions 1–8 from the affinity purification. *Sr*-Gal-3 was eluted in the last three fractions 6–7 as indicated by the arrow. Fraction 8 shows highly purified *Sr*-Gal-3.

3.4.1.6 Antibody recognition of *Sr*-Gal-3

To demonstrate the immunogenicity of the secreted *Sr*-Gal-3, recombinant *Sr*-Gal-3 was tested in ELISA using both, rat serum and human serum of individuals previously infected with *S. stercoralis*. In comparison Figure 3.4.1.6-*1* also shows the immunogenic properties of an iL3 extract. It was demonstrated that the rat and human sera recognise the recombinant protein and also the extract proteins, whereas the extract reveals a higher intensity.

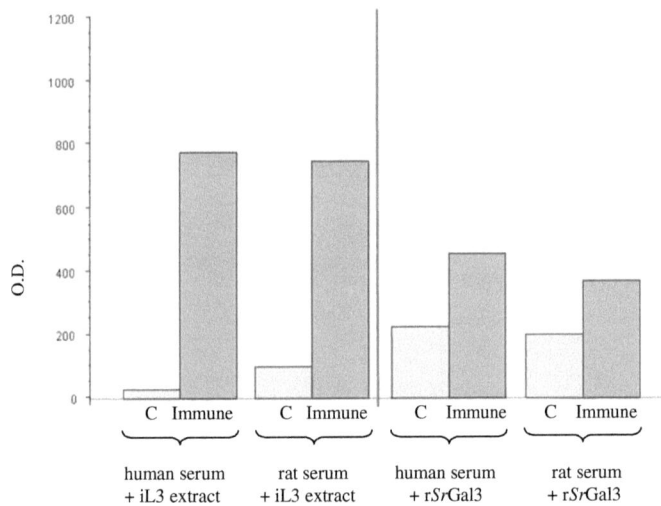

Figure 3.4.1.6-1 ELISA analysis shows the antibody recognition of recombinant galectin-3 (rSrGal3) and iL3 extract by sera from *S. ratti* infected rats (N=6) and *S. stercoralis*-infected humans (N=2). **C** - control serum from non-infected individuals, **Immune** – serum from infected individuals.

3.4.1.7 Sugar-binding assay of galectins

The capacity of galectins to bind carbohydrate structures containing β-galactosides can be used to determine the sugar-binding specificity of single galectins or protein mixtures. Herein the galectins contained in iL3 extracts were tested for the specificity for certain carbohydrate structures using carbohydrate arrays. The studies were kindly performed by Tim Horlacher from the Seeberger Glycomic research group at the ETH, Zurich. The principle of the assay includes the incubation of the proteins, e.g. iL3 extract, on the microarrays to allow them to bind to the exposed carbohydrates. Thereafter the unbound proteins are washed from the surface. In a second step antibodies from infected rat sera were bound to the attached proteins and detected with a secondary antibody (Horlacher, 2008; Figure 3.4.1.7-*1*)

Results

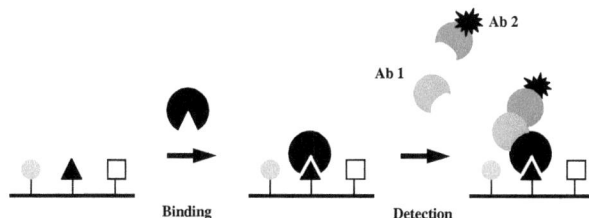

Figure 3.4.1.7-1 Principle for the conduction of carbohydrate microarrays. Binding of the protein to the arrayed sugars, binding of the serum antibodies (Ab 1) and the secondary antibody (Ab 2) and read out by a fluorescence scanner (from Horlacher, 2008; modified).

The results show that the extracts contain proteins that are able to bind to distinct β-galactoside structures. The subsequent incubation with sera from infected rats used to visualise the bound components demonstrates that the rats produced antibodies against these proteins. Figure 3.4.1.7-2 shows a carbohydrate microarray with the different binding intensities of the extract components. The corresponding carbohydrate structures bound to the microarray surface are listed in figure 3.4.1.7-3. It appears that, exept structures 8 and 11, the remaining carbohydrates, predominantly type 7, 12 and 13, are able to bind galectins in varying intensities.

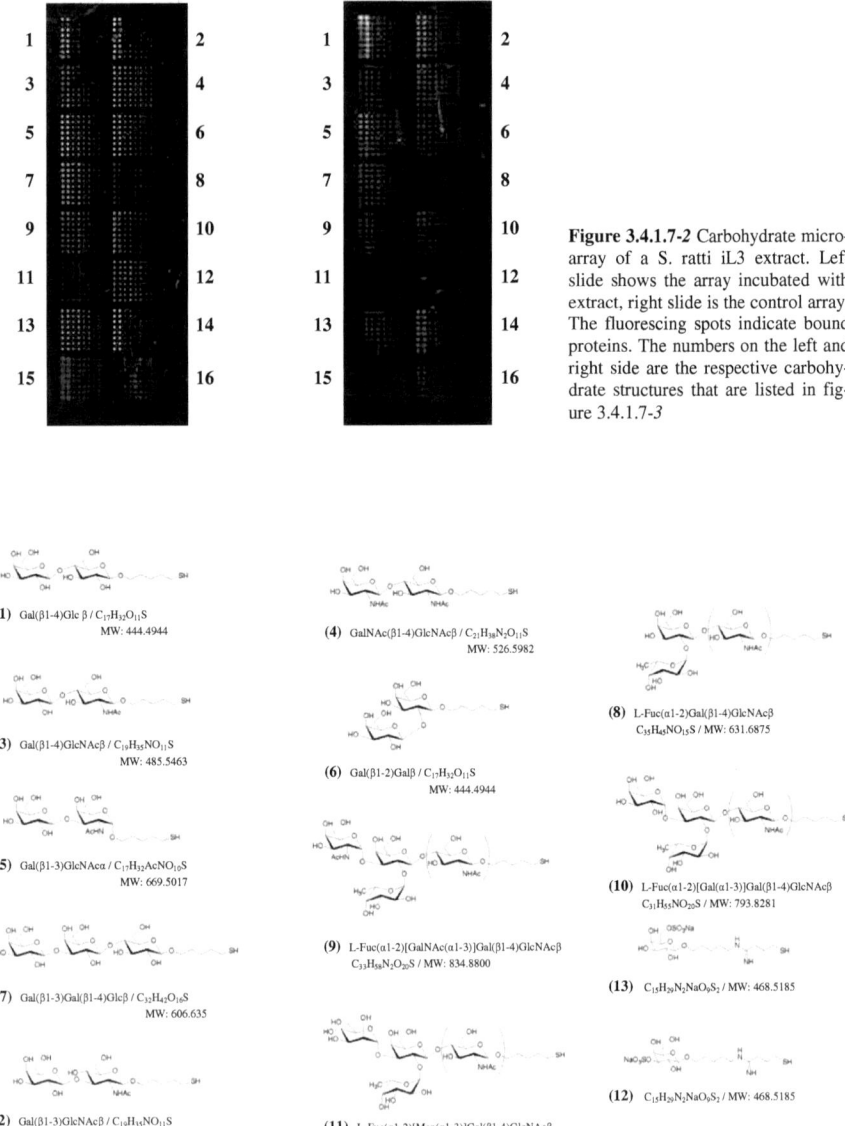

Figure 3.4.1.7-2 Carbohydrate microarray of a S. ratti iL3 extract. Left slide shows the array incubated with extract, right slide is the control array. The fluorescing spots indicate bound proteins. The numbers on the left and right side are the respective carbohydrate structures that are listed in figure 3.4.1.7-3

(1) Gal(β1-4)Glc β / $C_{17}H_{32}O_{11}S$
MW: 444.4944

(3) Gal(β1-4)GlcNAcβ / $C_{19}H_{35}NO_{11}S$
MW: 485.5463

(5) Gal(β1-3)GlcNAcα / $C_{17}H_{32}AcNO_{10}S$
MW: 669.5017

(7) Gal(β1-3)Gal(β1-4)Glcβ / $C_{32}H_{42}O_{16}S$
MW: 606.635

(2) Gal(β1-3)GlcNAcβ / $C_{19}H_{35}NO_{11}S$
MW: 485.5463

(4) GalNAc(β1-4)GlcNAcβ / $C_{21}H_{38}N_2O_{11}S$
MW: 526.5982

(6) Gal(β1-2)Galβ / $C_{17}H_{32}O_{11}S$
MW: 444.4944

(9) L-Fuc(α1-2)[GalNAc(α1-3)]Gal(β1-4)GlcNAcβ
$C_{32}H_{58}N_2O_{20}S$ / MW: 834.8800

(11) L-Fuc(α1-2)[Man(α1-3)]Gal(β1-4)GlcNAcβ
$C_{31}H_{55}NO_{20}S$ / MW: 793.8281

(8) L-Fuc(α1-2)Gal(β1-4)GlcNAcβ
$C_{15}H_{45}NO_{15}S$ / MW: 631.6875

(10) L-Fuc(α1-2)[Gal(α1-3)]Gal(β1-4)GlcNAcβ
$C_{31}H_{55}NO_{20}S$ / MW: 793.8281

(13) $C_{15}H_{20}N_2NaO_9S_2$ / MW: 468.5185

(12) $C_{15}H_{20}N_2NaO_9S_2$ / MW: 468.5185

Figure 3.4.1.7-3 Structures and molecular formulas of the carbohydrates that are covalently bound to the surface of the glass slides. The control spaces of structures 14 to 16 are not shown. Structure 14 is the Gal-linker, 15 is the Di-mannose-linker and 16 was treated with PBS meaning that no structure is bound to the surface.

3.4.2 Identification and analysis of a *S. ratti* prolyl oligopeptidase

In section 3.3.4.2 it was shown that E/S products from parasitic females contained distinct sequences with prolyl oligopeptidase (POP) similarities at a high abundance. It was therefore decided to further elucidate the sequences as a possible target for the treatment of *Strongyloides* infection.

3.4.2.1 Completion of the *Sr*-POP-1 sequence

By screening the EST database, additional to the two already known clusters SR01641 and SR03191, a third EST-cluster, SR03122, with homology to other prolyl oligopeptidases was found. The three EST protein sequences are 126, 189 and 209 amino acids long. By sequence comparison it was shown that SR01641 and SR03191 had 30 overlapping amino acids at their N- and C-terminal ends. Also SR03191 and SR03122 had overlapping sequences of 40 amino acids. Figure 3.4.2.1-*1* displays the linkage of the sequences resulting in a fragment with a total of 454 amino acids.

```
SR01641    YEYLENLQGKKTLEFVKNLNKISNKYLNQISIRSYIKRKIVQYYNYGKHSLFSKHGEYYYYTYKPPKKDH    70
SR03191    ----------------------------------------------------------------------
SR03122    ----------------------------------------------------------------------

SR01641    AILLRKYKYYYRGEVVFDVDKFDKTGKTSTKSISLPKNGKYIALLLSVNGSDWGTI--------------    126
SR03191    -----------------------KTSMKSISLPKNGKYIALLLSVNGSDWGTIRFMTNKGETLTNSL     44
SR03122    ----------------------------------------------------------------------

SR01641    ----------------------------------------------------------------------
SR03191    KNIKFTNMEFAYSGKGFFYSTFVNEKGQVVSNQKEKNVYHALLYHKMGRCQEDDIIIADYKEIDNMIILG   114
SR03122    ----------------------------------------------------------------------

SR01641    ----------------------------------------------------------------------
SR03191    SVSNDERYLFVYYYKGSSRENMIYYLNLSKFRRGKIHKKPKLKPLFTDFDGTYSIINTNCDELIVLTTKD   184
SR03122    ----------------------------------IHKKPKLKPLFTDFDGTYSIINTNCDELIVLTTKD    35

SR01641    ----------------------------------------------------------------------
SR03191    APTGK-----------------------------------------------------------------   189
SR03122    APTGKIIKMHIKEXKKWKTLIEADPKRKIKNVEAGGQKYLIVHYSENLKDRVYIYNKNNGKMITKLDLDS   105

SR01641    ----------------------------------------------------------------------
SR03191    ----------------------------------------------------------------------
SR03122    GSVVSISASPYYSRFFIKVSNQVIPQIIYTGNLMEMKHGKRKVTMRVIIKTTLYGIEKQNFVMKTEYYKS   175

SR01641    ----------------------------------------------------------------------
SR03191    ----------------------------------------------------------------------
SR03122    KDGTMIPMFIFHKKGIKLNGRNPVLLXXXXXXXX------------------------------------   209
```

Figure 3.4.2.1-*1* Alignment showing the overlapping N- and C-terminal ends of the three putative prolyl-oligopeptidase cluster sequences

For sequencing of the 5'-end parasitic female RNA was first reverse transcribed using the T7I oligo dT primer. Then PCRs were performed using POPr1f and POPr2f as forward and the T7II primer as reverse primers. The resulting PCR products (Figure 3.4.2.1.-2) were cloned into the pGEM-T Easy vector and sequenced using the M13 forward primer. The same cDNA was used for the investigation of the 3'-coding region. 3'-RACE PCR experiments were performed using the GeneRacer™ Kit according to the manufacturer's protocol. For sequencing the 3'-end the products were cloned into the pGEM-T Easy vector and sequenced using the M13 forward and gene specific reverse primers. By overlapping the previously known and the newly investigated sequences the full length sequence was obtained. Primers including the 3'- and the 5'-ends were designed and the full length fragment was captured by PCR. The result was confirmed by sequencing using the M13 forward-, the M13 reverse- and gene specific primers on the fragment that has previously cloned into the pGEM-T Easy vector. Figure 3.4.2.1-2 shows the intermediate PCR results and the full length PCR fragment. The resulting sequence was termed Sr-POP-1.

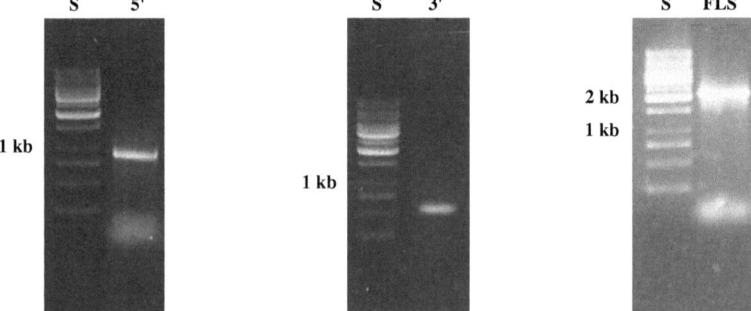

Figure 3.4.2.1-2 PCR results for the completion of the prolyl-oligopeptidase. Left side: 5'-end, Middle: 3'-end, Right side: Full Length Sequence (FLS) of the POP1

3.4.2.2 Mass spectrometric analysis of *Sr*-POP-1

After the POP was fully sequenced it was translated with Expasy translate tool and the resulting sequence was added to the search database in FASTA format. Using the ProteinPilot™ search engine, searches were performed again with the data files in which the EST cluster numbers SR01641, SR03191 and SR03122 have previously been found. The searches resulted in 25 peptides with a confidence score of 99 % and a sequence coverage of 32%. The *unused protein score* of 65.71 makes it the highest scoring protein among the sequences that were found for the parasitic females only. Table 3.4.2.2-*1* shows peptide information identified for the *S. ratti* POP1 sequence and figure 3.4.2.2-*1* shows the sequence coverage where the green peptides represent the identified sequences.

Table 3.4.2.2-*1* Peptides matching to *Sr*-POP-1 and having a confidence score (Conf.) of 99 %

Conf. %	Sequence	Modifications	Cleavages	ΔMass	Spectrum
99	CQEDDIIIADYK	Carbamidomethyl(C)@1		0.83465666	1.1.1.4382.1
99	EIDNMIILGSVSNDER	Oxidation(M)@5		0.86324477	1.1.1.4605.1
99	ENMIYYLNLSK			0.90008485	1.1.1.5464.1
99	ETVDVFAFIK			0.78791523	1.1.1.6359.1
99	FTNMEFAYSGK	Oxidation(M)@4		1.84060276	1.1.1.3395.1
99	GETLTNSLK			0.09741686	1.1.1.2731.1
99	HGEYYYYTYKPPK			1.55479634	1.1.1.3268.1
99	ILNECEKNELYFR	Carbamidomethyl(C)@5	missed K-N@7	0.89058173	1.1.1.4014.1
99	IVQYYNYGK			0.83453107	1.1.1.3205.1
99	KDHAILLR			-0.21634747	1.1.1.1427.1
99	KIVQYYNYGK		missed K-I@1	0.73956805	1.1.1.2882.1
99	KIWNHYEYLENLQGK	Deamidated(Q)@13	missed K-I@1	1.46544945	1.1.1.4228.1
99	KIWNHYEYLENLQGKK	Deamidated(N)@4	missed K-I@1; missed K-K@15	-0.71941835	1.1.1.3892.1
99	LDLDSGSVVSISASPY-YSR			0.5407356	1.1.1.5501.1
99	NELYFR			0.06971104	1.1.1.3416.1
99	NVYHALLYHK			0.43885049	1.1.1.3068.1
99	RPIQWPSTLITTGLNDDR			0.97232467	1.1.1.5616.1
99	SEYGDPEKEEDFNYLL-NYSPLNNLK	Deamidated(N)@13; Deamidated(N)@22		0.18803531	1.1.1.6536.1
99	VKETVDVFAFIK			0.72451389	1.1.1.5752.1
99	VSNQVIPQII-YTGNLMEMK			1.62301004	1.1.1.6524.1
99	YAAELYYTIQK			1.75869	1.1.1.4461.1
99	YLIVHYSENLK			1.8361516	1.1.1.3960.1
99	YLIVHYSENLKDR			0.78817564	1.1.1.3546.1
99	YLNQISIR			0.58469057	1.1.1.3783.1

```
MHYSLFLYYIFALVALLPAIDSKKSSEKRNKNSSKRPNLITPQPKILNECEKNELYFRNRYNSLIRINVTTYPIIERC
NTCSKIVFGKKIWNHYEYLENLQGKKTLEFVKNLNKISNKYLNQISIRGYIFKKIVQYYNYGKHSLFGKHGEYYYYTY
KPPKKDHAILLRKYKYYYRGEVVFDVDKFDKTGKTGTKSISLPKNGKYIALLLSVNGSDWGTIRFMINKGETLTNSLK
NIKFTNMEFAYSGKGFFYSTFVNEKGQVVSNQKEINVYHALLYHKMGRCQEDDIIIADYKEIDNMIILGSVSNDERYL
FVYYYKGSSRENMIYYLNLSKFRRGKIHKKPFLKPLFTDFDGTYSIINTNCDELIVLTTKDAPTGKIIKVNIKDAHKG
IKKWKTLIKADPKRKIKNVEAGGQKYLIVHYSENLKDRVYIYNKNNGKMITKLDLDSGSVVSISASPYYSRFIIKVSN
QVIPQIIYTGNLMEMKHGKRKVIMRVIIKTTLYGIEKQNFVMKTEYYKSKDGTMIPMFIFHKKGIKLNGRNPVLLEGY
GGFGVSFLPTFSSSNLMFVNHLNGIYVIACIRGGGEYGKKWHDAGKHLNKQNSFDDFIAAAEYLINEKYTNPSKLAIS
GSSNGGLLTAVVSQQRPELFGTVIIGVGVLDMIRYHNFTFGAAWKSEYGDPEKEEDFNYLLNYSPLNNLKMFKRPIQW
PSTLITTGLNDDRVVASHSLKYAAELYYTIQKGIRYQRNPXLVKVFDGQGHNGAITSIKRVKETVDVFAFIKETLNIK
WTYQVKN
```

Figure 3.4.2.2-1 Sequence coverage of the MS data obtained for the *Sr*-POP-1 full length sequence. Green peptides have a coverage of 99 and yellow peptides ≥50 - <99.

3.4.2.3 Sequence analysis of *Sr*-POP-1

The full length sequence was now further analysed. Figure 3.4.2.3-*1* shows the full length nucleotide sequence and includes the corresponding amino acid sequence. The open reading frame of *Sr*-POP-1 contains 2,364 nucleotides and the gene product covers 797 amino acids. The yellow nucleotide sequences are the previously unknown parts of *Sr*-POP-1. 278 nucleotides were missing on the 3'-end and 736 on the 5'-end. The domain structure of the POP was analysed as described in section 3.4.1.2.

A schematic description of the protein is shown in figure 3.4.2.3-*2*. The protein is carrying a signal peptide with the cleavage site between position 22 and 23 and the calculated molecular weight is 91 kDa (http://www.expasy.ch/cgi-bin/protparam). The signal peptide sequence is followed by several alternating potentially disordered globular regions. The domain characteristic for POPs is the Peptidase_S9_N region which is covered by residues 72 to 487 and the serine active site lies between residues 602 and 632 according to the Prosite database.

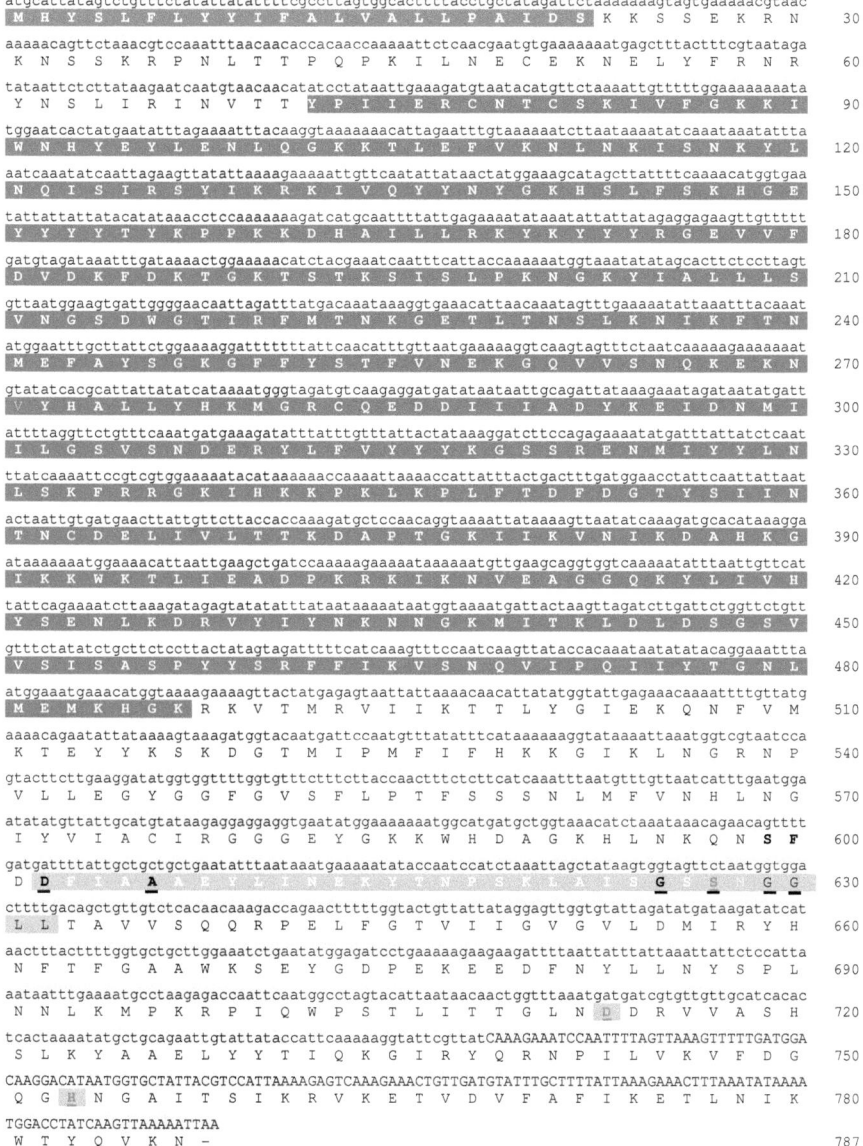

Figure 3.4.2.3-1 Full length nucleotide sequence of *Sr*-POP-1. The corresponding protein sequence is shown in single letter code. The previously unknown nucleotide sequences are yellow. The signal peptide is marked in red, the Peptidase_S9_N region blue and the serine active site light green. Black amino acids inside the serine active site are highly conserved. The red amino acids Ser627, Asp713 and His753 represent the catalytic triad.

Figure 3.4.2.3-*3* shows the sequence logo which compares the serine active site of 24 different POP enzymes. The six tall amino acid residues, where the heights of each letter in the logo position is proportional to the observed frequency of the corresponding amino acid in the alignment column, are also present in *Sr*-POP-1 and marked by black bars under the respective residues in the protein sequence (Figure 3.4.2.3-*1*). Residue 26 in the sequence logo is a serine that is located in position 627 of *Sr*-POP-1 and is part of the catalytic triad together with aspartic acid 713 and histidine 753. The residues of the catalytic triad are shown in red in figure 3.4.2.3-*1*. The conserved serine has been shown to be essential for the catalytic mechanism (Szeltner, 2008).

To locate the catalytic triad a three dimensional structure was generated using the Swiss Model server version 8.05 (Arnold, 2006; http://swissmodel.expasy.org/). For modelling the 3-D-structure a known structure of a POP from porcine brain was used as a template (Swiss Model template number: 1e5tA). As illustrated in figure 3.4.2.3-*4* the enzyme has a cylindrical shape and consists of two domains, a peptidase and a seven-bladed β-propeller. The catalytic triad is located in a large cavity at the interface of the two domains. The serine 627 (white) is found at the tip of a sharp turn and directly next to histidine 753 (violet), containing the catalytic imidazole group. The adjacent aspartic acid 713 (yellow) is connected to the histidine by formation of

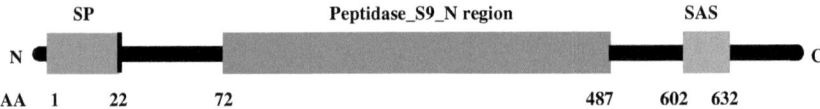

Figure 3.4.2.3-*2* Domain structure of the *Sr*-POP-1. **AA** – amino acid, **SP** – signal peptide, **SAS** – serine active site

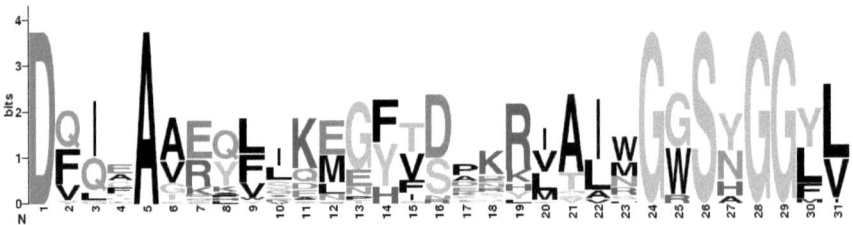

Figure 3.4.2.3-*3* Sequence logo of the serine active site generated from multiple sequence alignments of 24 different POP active sites. The total height of a logo position depends on the degree of conservation in the corresponding multiple sequence alignment column. Very conserved alignment columns produce high logo positions. The highest amino acids are marked by black bars under the respective residues in the sequence shown in figure 3.4.2.3-*1* (source http://us.expasy.org/cgi-bin/prosite).

a hydrogen bond between one of the two oxygen atoms of the carboxylate group and the N-hydrogen of the imidazole ring of histidine.

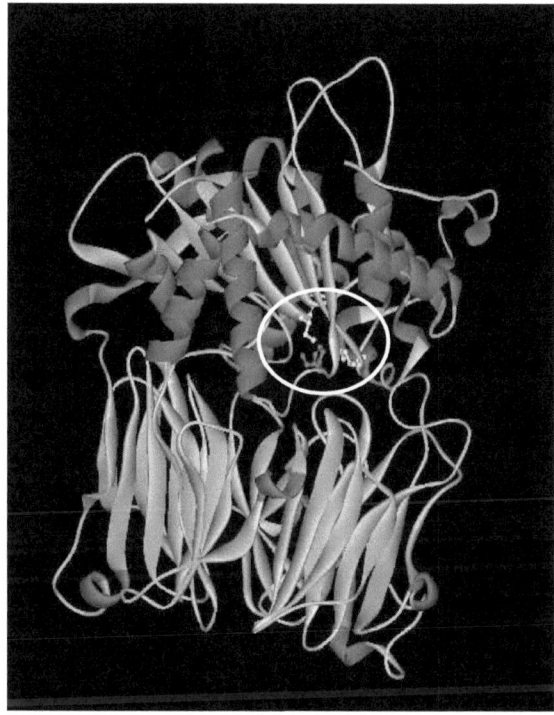

Figure 3.4.2.3-4 Ribbon diagram of Sr-POP-1. The peptidase domain is located in the upper region, the β-propeller domain in the lower region.
a) light blue - β-propeller structures
b) red – helical structures
c) white – Ser627
d) violet – His753
e) yellow – Asp713

3.4.2.4 Phylogenetic analysis of *Sr*-POP-1

Sr-POP-1 represents a newly identified and stage-specific protein of the parasitic nematode *S. ratti*. To compare the protein to sequences from other nematodes and vertebrates a BLAST search was performed at the NCBI protein database. The highest scoring and only BLAST hit referring to another nematode was a POP family protein from *B. malayi*. No significant alignment was found for the non-parasitic nematode *C. elegans* which is unusual because a majority of proteins are commonly shared between *C. elegans* and other nematode species. The second BLAST hit is represented by a POP family member from *R. norvegicus* followed by other vertebrate and non vertebrate species. Selected sequences were aligned using ClustalW2 and the sequence identities of both whole enzyme and the C-terminal ends containing the catalytic regions were compared (table 3.4.2.4-*1*). The sequence alignment of the C-terminal ends is shown in figure 3.4.2.4-*1*. The homologies of the proteins range between 34–37% with a shift to 47–52% if only the C-terminal ends are compared. The sequence alignment of the active sites shows the highly conserved regions including the previously mentioned six amino acids being conserved in 24 different POPs. Also the catalytic triad is present in all of the listed sequences. The multiple sequence alignment peptidase domains were used to construct a phylogenetic tree where also the sequence of the protozoan parasite *T. cruzi* was used in addition (table 3.4.2.4-*2*). *S. ratti* and *B. malayi* as the only representatives of parasitic nematodes are allocated in the same cluster and show a higher similarity to POP family enzymes from *T. cruzi* and *D. melanogaster*.

Table 3.4.2.4-*1* Homologies of *Sr*-POP-1 with related proteins from selected species

Organism	% Identity with	
	whole protein	C-terminal region (incl. catalytic domain)
B. malayi	37	52
R. norvegicus	36	49
H. sapiens	35	48
M. musculus	36	50
D. melanogaster	34	47
S. scrofa	35	48

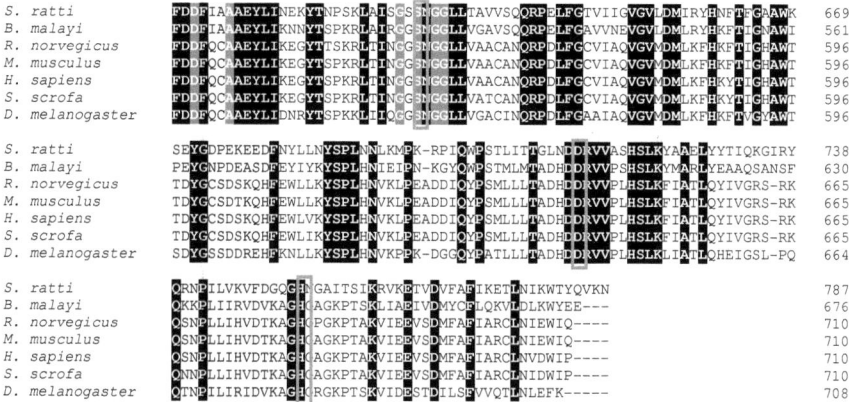

Figure 3.4.2.4-1 Multiple sequence alignment of the C-terminal ends of different POPs. Homologous residues have black backgrounds. Highly conserved residues have red backgrounds. Residues of the catalytic triad are in green boxes. Accession numbers: *B. malayi* - XP_001894227, *R. norvegicus* - NP_112614, *M. musculus* - NP_035286, *H. sapiens* - CAA52605, *S. scrofa* - NP_001004050, *D. melanogaster* - NP_609397.

Figure 3.4.2.4-2 Phylogenetic tree of selected POPs. The length of the branches is proportional to the evolutionary distance of the species. The POPs from *B. malayi* and *S. ratti* are in the same cluster.

3.4.2.5 Inhibition of the *Sr*-POP-1 enzyme activity

Based on the finding that *Sr*-POP-1 represents an abundant secreted protein specific for the parasitic female stage it was decided to investigate whether POP inhibitors show an effect when added to *in vitro* cultures of parasitic females. POP has gained pharmaceutical interest, since its inhibitors have been shown to have anti-amnesic properties in rat. Since then several potent POP inhibitors have been synthesised and it was shown that they are capable of increasing the brain levels of several neuropeptides (Atack, 1991), to reverse scopolamine-induced amnesia in rats (Toide, 1996) and to improve cognition in old rats (Toide, 1997) and MPTP-treated monkeys

(Schneider, 2002). The effects of POP inhibitors on parasitic nematodes, however, have not been investigated before, since until today, POP's have not been implicated to play a role in the metabolism of nematodes. Therefore several POP inhibitors were requested and kindly supplied by Elina M. Jarho from the Department of Pharmaceutical Chemistry, University of Kuopio, Finland. The tested compounds are presented in figure 3.4.2.5-*1* and have previously been shown to have inhibitory potential on rat POP (Venäläinen, 2006).

0.5 M sterile stock solutions of the four compounds were added to cultures of freshly prepared parasitic females shown to abundantly express the peptidase. The compounds were tested at final concentrations of 1, 2, 3, 5, 7.5 and 10 mM. The cultures were incubated at 36°C and the movement of the females was tested under the microscope at different time points. It was observed that the treated worms all showed a decreased degree of movement in comparison with the non-treated control group starting within the first 30 minutes after addition of the POP inhibitors. With increasing inhibitor concentrations the worms were moving slower and showed at both higher concentrations and longer incubation times no movement at all. Non-moving and slowly moving worms were counted. After 18 hours the culture medium was replaced with fresh medium not containing the inhibitors to verify that the motility does not recur when the female

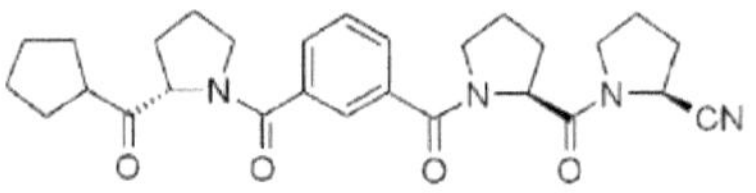

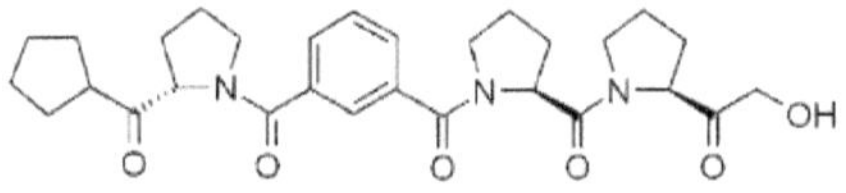

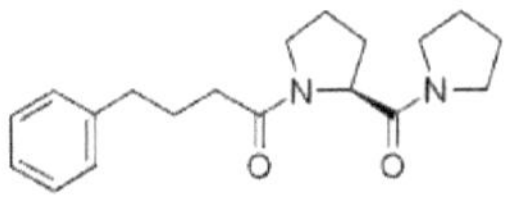

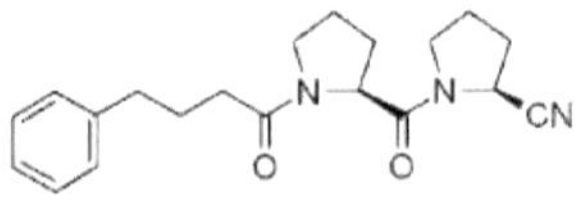

Figure 3.4.2.5-*1* The prolyl oligopeptidase inhibitors used in the study:
1A - isophthalic acid 2(S)-(cyclopentanecarbonyl)pyrrolidine-L-prolyl-2(S)-cyanopyrrolidine amide;
1B - isophthalic acid 2(S)-(cyclopentanecarbonyl)pyrrolidine-L-prolyl-2(S)-(hydroxyacetyl)pyrrolidine amide;
2A - 4-phenylbutanoyl-L-prolyl-pyrrolidine;
2B - 4-phenylbutanoyl-L-prolyl-2(S)-cyanopyrrolidine
(Venäläinen, 2006)

worms are not any longer exposed to the substances. The worms did not start to move again and were therefore considered as dead. As shown in figure 3.4.2.5-2 a dose dependency can be observed resulting in the maximum number of dead worms at 10 mM for every compound. The compounds 1A and 1B are less effective then 2A and 2B.

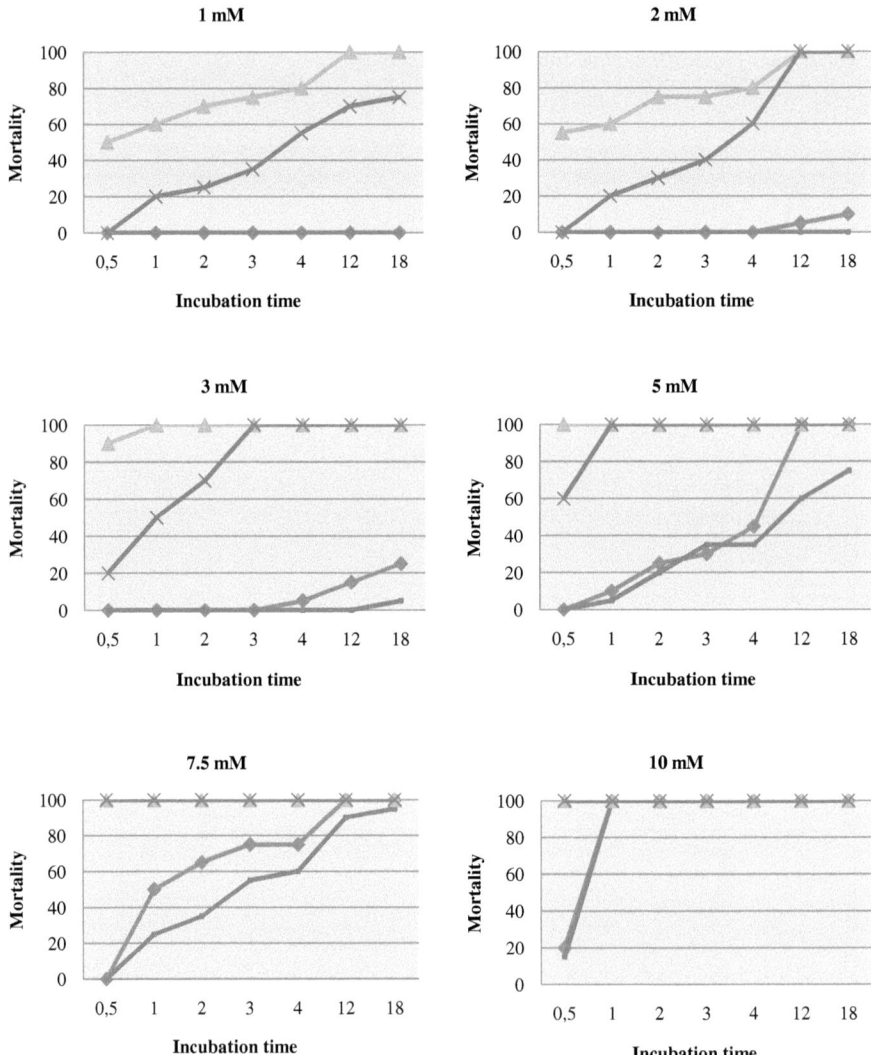

Figure 3.4.2.5-2 Effect of different POP inhibitors during *in vitro* culture of *S. ratti* parasitic female worms. Single graphs show different concentrations of compounds 1A ( ), 1B ( ), 2A ( ) and 2B ( ). The mortality represents the percentage of worms that do not show any movements at a certain time point.

4 Discussion

E/S products secreted by cells and organisms play pivotal biological roles across a wide range of parasitic organisms and can represent 8 ± 20% of their proteomes (Greenbaum, 2001). Being the primary interface between the parasite and the host, the E/S components include proteins known to be involved in biological processes representing cell migration, cell adhesion, cell-cell communication, proliferation, differentiation, morphogenesis and the regulation of immune responses (Maizels, 2003). Many proteins identified in the presented analysis reflect some of the aforementioned processes and also display a variety of further protein families and proteins with unknown function. Here for the first time the composition of these complex mixtures from different developmental stages of *S. ratti* was analysed. To address the question whether proteins are present in the culture supernatant due to secretion or leakage of dead or damaged individuals, cycloheximide (CHX), an inhibitor of protein translation, was added to cultures of iL3. In addition, crude extracts were prepared from infective larvae, parasitic females and free-living stages and submitted to mass spectrometric analysis. Therefore it was possible to verify protein secretion by comparing the culture supernatants versus CHX cultures and extracts. By choosing the technology of tandem mass spectrometry it was possible to fulfil the aims of this work described herein. It was emphasised to identify proteins that are abundantly or differentially produced by variable developmental stages occurring in the life cycle of the parasitic nematode *S. ratti*. In a second step specific proteins that might be essential for establishment and containment of parasitism and proteins with putative modulatory effects on the host immune system were chosen for further investigation by molecular biology methods.

4.1 Comparison of results with published EST data from *Strongyloides* ssp.

One of the problems that had to be resolved when performing proteomics with *S. ratti* was the fact that its genome was not being sequenced at the beginning of this project. In the meantime the Wellcome Trust Sanger Institute has started the process of sequencing the *S. ratti* genome as reference genome and the subsequent whole genome shotgun sequencing of the *S. stercoralis* genome planned. Since the sequencing project is not completed until today, known EST sequence databases from *S. ratti* and *S. stercoralis* had to be utilised for searching the mass spectrometry data. In total 14,701 ESTs from *S. ratti* and 11,335 ESTs from *S. stercoralis* are known until today. These ESTs were overlapped and the resulting contigs were grouped into 4,152 clusters from *S. ratti* (Thompson, 2005) and 3,311 clusters from *S. stercoralis*, respectively. A custom search database was created by combining both *Strongyloides* EST cluster datasets together with protein sequences from *C. elegans*, *C. briggsae* and other parasitic nematodes. Offering the opportunity to search in amino acid substitution mode the ProteinPilot™ search engine was chosen for the evaluation of the mass spectrometry data instead of the Mascot search engine. Thompson *et al.* assumed that the genome of *S. ratti* codes for approximately 22,239 proteins as does *C. elegans*. Following this assumption the 1,081 identified proteins of the presented work cover about 4.9% of the total number of proteins estimated to be coded by *S. ratti*. However, the mentioned 4,152 and 3,311 clusters represent only 18.7% and 14.9% two *Strongyloides* genomes. Therefore it can be stated that once genomic data from *S. ratti* becomes available, the total number of proteins that can be analysed in the presented study data will rise since the searches were performed against 18.7% and 14.9% of the *S. ratti* and *S. stercoralis* genomes.

In the *S. ratti* EST analysis published in 2005 the clusters were ranked by their number of EST members and the top 30 clusters, that represented 38% of the total ESTs obtained were investigated in detail. Only four out of the 30 clusters were also identified in the proteomic analysis presented here. One of these clusters - SR00984 - was already presented in the result section listing the 25 highest scoring proteins found exclusively in samples from parasitic females. This cluster has a significant BLAST alignment to a small heat-shock protein from *T. spiralis*, a parasitic nematode of boar, whereas in the published data relating to the EST analysis project the BLAST alignment was significant to heat-shock protein 17 from *C. elegans*. This different finding can be explained through the latter publication date of the *T. spiralis* protein sequence. Sequence alignment of the *S. ratti* and *T. spiralis* sequences shows a similarity of 35% with the *T. spiralis* sequence being five amino acids longer. It is likely that the *S. ratti* sequence, that is composed of 160 amino acids represents the complete protein since direct translation of the corresponding cluster nucleotide reveals the start and the stop codon. For the recombinant *T. spiralis* heat-shock protein it was reported that it possesses chaperone activity to inhibit the heat-induced aggregation of citrate synthase. Also it was found that the protein expression is thermally induced and developmentally regulated, mainly in the mature muscle larvae (Wu, 2007). The authors suggested that the heat-shock protein likely plays a role in enhancing the survival of the *T. spiralis* muscle larvae under conditions of chemical and physical stress, as well as in the development of larvae. By antibody recognition it was shown that the heat-shock protein 10 is strongly immunogenic. For the *S. ratti* cluster SR00984 this work supports the finding of

Thompson *et al.* (2005) namely that the protein is expressed stage-specific. In the presented study the named cluster was shown to be specific for parasitic females. In the EST analysis the 81 ESTs forming the cluster SR00984 were also found only in the library generated from parasitic female material. In addition a comparative microarray analysis of *S. ratti* also showed that various orthologues of *C. elegans* heat-shock gene, *hsp-17*, were upregulated in parasitic females (Evans, 2008). The hypothesis made by other researchers that the heat-shock proteins may be used for monitoring changes in an organism's environment might be also an explanation for *S. ratti*. At least it is likely to be important for parasitic nematodes because they move between different host species (Thompson, 2001) or for *Strongyloides* spp. between host and non-host environments.

The second protein representing one of the 30 top EST clusters is SR01068, a ribosomal protein shown to be expressed at a significantly higher level in free-living stage libraries compared with parasitic female libraries. The difference compared to the previously discussed cluster SR00984, that was found only in the parasitic female samples in both studies of the EST analysis presented by Thompson *et al.* and the proteomic analysis presented here, is that the cluster SR01068 is found in the free-living as well as in the parasitic female libraries albeit with a lower representation. In the proteomic analysis SR01068 was found in E/S products from free-living and parasitic female stages and extract samples also from the iL3. BLAST search showed a significant alignment with ribosomal protein 1 (rpl-1) from *C. elegans*, a molecule involved in protein biosynthesis. A high number of ribosomal proteins was also identified in E/S products from parasitic females (see table 3a). Nagaraj *et al.* used ESTs from 39 economically important and disease-causing parasitic nematodes of humans for the large-scale identification and analysis of E/S products. In that study ribosomal proteins from *Haemonchus contortus*, *Trichuris muris*, *Globodera rostochiensis* and *B. malayi* were classified as E/S proteins (Nagaraj, 2008) underlining the likelihood of ribosomal proteins being present in *S. ratti* culture supernatants as well as extracts. Since the method applied for this study is not capable of observing expression levels for singular proteins during the life cycle no prediction can be made concerning the role of ribosomal proteins observed in one or more stages. However Thompson *et al.* by performing a microarray analysis of gene expression in the free-living stages of *S. ratti* found SR01068 to be 2.08-fold higher expressed in L1 stages compared to iL3. They hypothesised that the higher occurrence of ribosomal proteins in L1 can be explained by the need of this stage to undertake substantial protein synthesis for the growth and development into the L2 stages (Thompson, 2006).

A similar explanation can be applied for the high occurrence of ribosomal proteins found in parasitic females, being stages that have to reproduce and produce eggs at high numbers.

EST cluster SR00605 was identified in supernatants and extracts from all stages as well as in control samples supplemented with CHX. It represents the third protein that is listed among the 30 largest clusters in the EST analysis of the life cycle of *S. ratti*. In the EST analysis the BLAST search showed an alignment with the *Ostertagia ostertagia* F7 E/S product whereas in this work the highest scoring protein is an *Onchocerca. volvulus* protein termed S1 protein. Using motif prediction tools shows that both proteins as well as SR00605 are likely to be members of the fatty acid and retinol binding proteins, a group of nematode specific proteins thought to be involved in complex host-parasite interactions (Basavaraju, 2003). Assuming that the 180 amino acid EST transcript represents the full length sequence a signal peptide analysis shows that the protein is carrying a signal peptide and is thus likely to be secreted. However, the fact that the protein is also found in samples supplemented with CHX leads to the assumption that the protein is highly abundant. Thompson *et al.* also stated that the high level of representation of the top 30 clusters attests to their likely biological significance in the life of *S. ratti* (Thompson, 2005). Using the ProteinPilot™ search engine revealed a second highly similar cluster in the EST database – SR01075. By directly comparing the sequences using sequence alignment tools it was shown that SR01075 is only two amino acids longer and has 95% homology to SR00605. Using the nemaBLAST search tool on www.nematode.net against the *S. stercoralis* EST database a matching cluster sequence – SS00108 - was identified. Direct comparison of the *S. ratti* and *S. stercoralis* using sequence alignment tools showed that SS00108 is lacking the N-terminal end and has 85% homology to *S. ratti*. Together with the fact that retinol-binding protein are thought to be involved in host parasite interactions this leads to the conclusion that SR00605 and SR01075 are interesting candidates for further investigation in the future.

SR01070 is the fourth cluster number that was found and represents one of the 30 largest EST clusters in *S. ratti*. In the presented work SR01070 can be found in table 12a listing the proteins that were analysed in extracts from the free-living stages. In contrast this cluster had no significant difference in its stage-specific expression profile in the EST analysis. In a BLAST search the cluster aligns to a *C. elegans* ABC transporter family protein and in the three dimensional map of *C. elegans* gene expression is found in, so called, mountains of gene expression that are enriched for germline expressed genes (Thompson, 2005). ABC transporters are a family of membrane proteins that share an ATP-binding cassette and can actively transport specific substrates through cell membranes. Even though Thompson *at al.* hypothesised that the large clusters reflect the likely biological significance in the life of *S. ratti* it remains unknown why 26 of these clusters were not identified in E/S products or extracts of either one of these stages. For those proteins that were overrepresented in the EST libraries from L1 and L2 stages the explanation could be that the amount of material was insufficient for a mass spectrometric analysis. The samples from the free-living stages contained a mixture of females, males, various larval stages and eggs. Due to the much larger size of the free-living adults it might be that the L1 and L2 stages were underrepresented. In this case it should be possible to identify the respective proteins in samples containing a pure suspension of the L1 or L2 stages.

In another work a microarray consisting of 2,227 putative genes was used to identify genes likely to play a key role in the parasitic life of *S. ratti* (Thompson, 2008). In the published data the microarray was probed with cDNA prepared from parasites subject to low or high immune pressures. Parasitic females under low immune response were harvested six days and females under high immune pressure were harvested 15 days post-infection. Comparison of the proteomic data with the microarray results shows that some of the previously termed putative genes were also present in the samples prepared from different *S. ratti* stages as shown in table 4.1-*1*. Interestingly the previously discussed cluster SR00984 which relates to a heat-shock protein 17 occurs again in this list in parasites under low immune pressure. As shown in table 10a the cluster number SR04440 aligns to the prolyl oligopeptidase sequence that was published online as a result of this work. The E-value of $2e^{-28}$ however shows that the fragment is not homologous to the sequence. Thus there must be another prolyl oligopeptidase protein beside the sequence published in this work. The fact that it is expressed in parasites under high immune response makes it another interesting candidate to study in the future. Proteins that also warrant further investigation are the clusters SR02091, SR01943 and SR00852 because BLAST searches show no significant alignment or alignments to hypothetical proteins.

Another finding is that just because proteins were found in one or more stage-specific samples they are not necessarily at the same time expressed in the respective stages only. They can also be expressed in other stages and were not detected by mass spectrometry due to underrepresentation. To finally prove that proteins are expressed at higher levels in certain stages it should be verified either by a regular PCR that shows the exclusive expression of a protein, as for example it has been shown for the astacin or the POP in the results section, or by real time

PCR that shows the higher or lower expression of proteins when directly comparing certain stages.

Table 4.1-*1* The 14 clusters that were found in different protein samples and also have a significantly higher expression in parasitic females subject to low or high immune responses (explained in the text). The numbers in the `table` column show the numbers of the protein lists that can be found in section 7 - Appendices – of the presented work.

	Cluster number	BLAST result	Table	Sample type(s)
Low immune pressure	SR00950	Activated protein kinase C receptor	1a	All stages E/S
	SR00984	Small heat-shock protein	3a	Only parasitic females E/S
	SR00843	T complex protein 1, zeta subunit	12a	Only free-living stages extracts
	SR00990	Peptidyl-prolyl cis-trans isomerase	1a	All stages E/S
	SR00756	ADP ribosylation factor 79F	1a	All stages E/S
	SR03324	NADH ubiquinone oxidoreductase	12a	Only free-living stages extracts
High immune pressure	SR02091	Hypothetical protein CBG22129	5a	Only free-living stages E/S
	SR04440	Prolyl oligopeptidase	10a	Only parasitic females extracts
	SR01943	Novel protein similar to COG3	2a	Only iL3 E/S
	SR00941	Aconitase family member	1a	All stages E/S
	SR00383	Propionyl Coenzyme A Carboxylase	2a	Only iL3 E/S
	SR00952	Heat-shock protein 90	1a	All stages E/S
	SR01051	Nucleoside di-phosphate kinase	1a	All stages E/S
	SR00852	Hypothetical protein CBG20301	12a	Only free-living stages extracts

4.2 Comparison with data from other parasites

A variety of proteins have been identified in both E/S products and extracts that are homologous to proteins from other nematode species. Some of these proteins have been identified in other species and have been implicated to be involved either in the containment of parasitism or in the suppression or induction of host immune responses. For *S. ratti* it has been shown that rats form a marked immunity against a challenge infection when immunised with E/S products of adult worms. The E/S products were injected directly into the lumen of the small intestine and led to a reduction of worm burden and faecal egg output compared with non-immunised rats (Mimori, 1987). For *S. ratti* however it has not yet been shown which proteins might lead to a protective immunity, whereas for other nematodes various proteins have been found to induce an immune response in their host species. In the following two representative *S. ratti* proteins, tropomyosin and astacin, are discussed and set into relation to homologous proteins from other parasitic nematode species.

Proteins that showed similarities to structural proteins in BLAST searches such as actin, profilin or myosin were detected in various samples from all stages. Also tropomyosin, a fibrillar protein involved in the contraction of muscle cells, was found in samples from all stages – free-living, parasitic females and iL3. The *S. stercoralis* sequence SS01430 represents the tropomyosin and can be found in table 1a. The ProteinPilot™ search engine led to the identification of further cluster numbers, e.g. SR00241, SS00615 and SS01171, with a lower *Unused Protein score*. Since these cluster numbers only represent a part of the identified SS01430 peptides they are not listed in the appended tables but also represent partial tropomyosin sequences. Tropomyosin family proteins were also found in secretions of adult *B. malayi* stages (Hewitson, 2008), in larval stages of *T. circumcincta* (Craig, 2006) and in *S. mansoni* cercarial secretions (Knudsen, 2005). Hartmann *et al.* (2006) showed that a partial protection against *A. viteae* challenge with L3 can be seen in jirds, when immunised with recombinant tropomyosin in conjunction with the Th$_1$-inducing adjuvant STP (a mixture of sqalane, tween and the polymer pleuronic). By examining the humoral response to tropomyosin in vaccinated rodents and showing that IgG and IgE antibodies react with the *O. volvulus* tropomyosin epitope Jenkins *et al.* (1998) suggested that it might also play a role in host-protective responses in onchocerciasis. In filarial nematodes tropomyosin is not only present in muscles but was also found in the cuticle of microfilariae and L3 of *O. volvulus*. This is supported by the fact that this protein was also detected in samples of worms treated with CHX.

As shown in section 3.2 a protease with high activity is produced by iL3. The overlay of the protease assay showed that the size of the enzyme is 33 kDa. The comparative PCR analysis presented in section 3.3.2 and the mass spectrometry results confirmed the stage specificity (see table 2a, page 126) and identified the protein to be an astacin-like metalloprotease. The designation SR11111 relates to a full length sequence that was identified in our laboratory based on a homologous *S. stercoralis* sequence (Borchert, 2006). The elevated expression of this enzyme in *S. ratti* and *S. stercoralis* (Gomez Gallego, 2005) in iL3 underlines its putative role to facilitate skin penetration at the initiation of infection. Borchert *et al.* (2007) showed that a homologous *O. volvulus* astacin is exposed to the host immune system and hypothesised that it may be a candidate for intervention strategies in filarial infections since another Astacin homologue is part of a promising hookworm vaccine. Besides the cluster SR11111 and its *S. stercoralis* homologue

SS01564 two further EST clusters, SR03587 and SR02663, that also showed similarities to astacins, were identified in E/S products from parasitic females. This indicates that *S. ratti* just as *C. elegans* exhibits an astacin family of which different members are expressed stage specific.

4.3 *S. ratti* galectins

4.3.1 Role of galectins in immune responses

Lectins sharing both their preference for binding β-galactosides and significant sequence similarity in the carbohydrate recognition domain (CRD) are commonly termed galectins (Barondes, 1994). Each galectin contains one or two highly conserved carbohydrate recognition domains (CRDs) made up of about 135 amino acid residues (Figure 4.3.1-*1*). The prototype galectins contain a single CRD which can form two-fold symmetric homodimers through interactions between the hydrophobic β-strands of each subunit. Thus, under normal conditions these lectins are bivalent molecules. The tandem galectins contain two homologous CRD domains separated by a proline-glycine-enriched linker peptide of about 20-30 amino acids. The chimera type galectins contain one CRD and a domain of about 110-130 residues, depending on species, that includes multiple repeat sequences rich in proline, tyrosine, glycine and glutamine (Hughes, 1999). In humans, 15 galectins have been described within the galectin family. In C. elegans there are 11 galectins, *Ce*Lec1 – *Ce*Lec11.

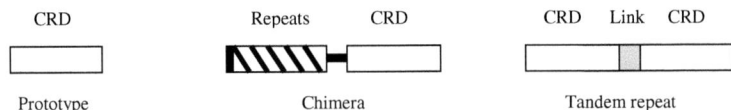

Figure 4.3.1-*1* Galectin structures. The homologous carbohydrate recognition domain (CRD) present in all galectins, the unique glycine-thyrosine-glutamine-proline-rich repeat sequence of the chimera type galectins and the link regions of tandem CRD galectins are shown. (Hughes, 1999)

Various biological roles of human galectins have been proposed, for example in regulation of immunity and inflammation, progression of cancer and in specific developmental processes (Leffler, 2004). In the past galectins and helminth infections have rarely been associated with each other but new data from *S. ratti* as well as from other parasitic nematodes showed high abundance of different galectins in the parasitic stages. The possible involvement of galectins in parasite infections becomes obvious when considering two further aspects. Firstly, galectins are involved in parasite infection and allergic inflammation, two very different but immunologically linked phenomena. Secondly, the hygiene hypothesis which suggests that allergic responses represent a misdirected activation of the arm of the immune system responsible for parasite attrition and that parasite infection may prevent the development of some allergic conditions (Yazdanbakhsh, 2004).

Host galectins are implicated in both establishing and combating parasite infections. The first line of recognition in an immune response is the parasites surface which is in general highly glycosylated and thus offers potential galectin binding sites. In this respect, there are examples of host galectins binding directly to glycoconjugates on the surface of parasites, leading to both positive and negative regulation of host immunity. For the protozoal parasite *Leishmania major* for example it has been shown that host galectin-9 binds specifically to the parasites surface lipophosphoglycans and thus is assisting in parasite binding to macrophages, promoting the cell invasion and finally facilitating infection (Pelletier, 2003). Human galectin-10 is present in eosinophils and basophils. The only other vertebrate galectin showing such restricted expression is the ovine galectin-14 that has recently been shown to be present only in ovine eosinophils (Young, 2008). Mucus collected from both lung and stomach of sheep with induced tissue eosinophilia from either allergen exposure with house dust mite or parasite infection with *H. contortus*, a parasitic nematode of ruminants, contains large quantities of galectin-14. This indicates that host galectins play a role during the immune response to parasitic infection.

On the other hand, the specific functions of the parasites galectins` is not known (Greehalgh, 2000), regardless of the variety that have been identified in helminth parasites so far. However, it has been shown that parasite galectins are involved in host-parasite interactions. For example a galectin from *O. volvulus* was recognised by sera from the majority of filaria-infected patients and was able to bind IgE (Klion, 1994). Furthermore two tandem-repeat type galectins have recently been discovered in a proteomic analysis of secretions from adult *B. malayi* stages (Hewitson, 2008). Johnston *et al.* (2009) stated that those galectins bind galactose containing glycoconjugates and may protect adult stages from the host's eosinophil and neutrophil mediated damage. A vaccination of goats with recombinant galectin antigen induced partial protection against *H. contortus* infection (Yanming, 2007). In another study it was shown that *H. contortus* produces a not further identified galectin, or possibly a mixture of galectins, which have potent chemokinetic activity for ovine eosinophils *in vitro* (Turner, 2008). This is interesting because experimental helminth infections have shown that eosinophils accumulate in the gastrointestinal tract, where it is thought that they help eliminate the parasite and for *S. stercoralis* infection eosinophil chemotaxis may have a central role in immunity (Mir, 2006). For the parasitic nematode *Dirofilaria immitis* it has been shown that in atopic individuals, resident in an area of canine endemia, the specific IgE response is stimulated mainly by two molecules, one of them being a member of the galectin family (Pou-Barreto, 2008).

4.3.2 Galectins identified in *S. ratti* E/S products and extracts

As reported in the result section 3.4.1 the proteomic analysis and the subsequent screening of *S. ratti* and *S. stercoralis* EST databases led to the identification of seven different galectins. The assigned names and the corresponding Strongyloides EST clusters are shown in table 4.3-*1*. The resulting sequences were blasted and named according to their closest relationship with *C. elegans* galectins. Four of these sequences, *Sr*-Gal-1, -2, -3 and -5, were found in the supernatants and three of the sequences, *Sr*-Gal-11, -21 and -22, remain hypothetical since the proteins were not found. However, PCR experiments with primers for all seven galectins showed the presence in cDNA of iL3. Using the nematode SL-1 primer and RACE-PCR it was possible to obtain the full length sequences of four galectins, *Sr*-Gal-1, -2, -3 and -22, showing that unlike the other galectins *Sr*-Gal-22 is composed of only one CRD. The attempts to obtain the full length sequences of *Sr*-Gal-5, -11 and -21 were not successful. To capture the missing fragments experiments should either be repeated using different primer sets or as soon as the *S. ratti* genome becomes available it will be possible to compose the sequences by BLAST search and multiple sequence alignment.

Table 4.3-*1* Seven *S. ratti* galectin sequences and the corresponding cluster numbers

Assigned Name	*S. ratti* EST Cluster	*S. stercoralis* EST Cluster	Homology
Sr-Gal-1	SR00627	SS00732	96
Sr-Gal-2	SRX0840 (SR05051)	SS00840	96 (93)
Sr-Gal-3	SR00900	SS00593	90
Sr-Gal-5	SR00857	SS02629	83
Sr-Gal-11	SR00257	SS01127	94
Sr-Gal-21	SR00838	/	/
Sr-Gal-22	SRX1190	SS01190	90

Discussion

An explanation for the identification of only four of the galectins might be that *Sr*-Gal-11, -21 and -22 are not being secreted. The secretion mechanism for galectins is ectocytosis in which cytosolic proteins concentrate in the cytoplasm underlying plasma membrane domains. The membrane then forms protrusions („blebs') including the previously formed protein aggregates. The blebs finally detach from the plasma membrane and are released as extracellular vesicles from which soluble proteins are released (also see section 1.7). Then, however, it should be possible to identify them in the extract samples but also the extracts revealed only sequences of *Sr*-Gal-1, -2, -3 and -5. Supposed that all mRNAs of the galectins are actually being transcribed the only remaining reason would be the underrepresentation of *Sr*-Gal-11, -21 and -22 among the vast number of different proteins in the samples. This opinion led to the attempts to enrich exclusively the galectins using lactose affinity separation, a method in which galectins are bound to lactose molecules. The lactose molecules are themselves covalently bound to agarose beads. After the separation process the galectins can be eluted by treating the beads with a lactose solution. This method was previously applied successfully to extracts of *H. contortus* iL3 (Turner, 2008). Since the preparation of sufficient amounts of E/S products is too time consuming for performing preliminary tests with this newly established method in the laboratory it was decided to use iL3 extracts instead. The test presented in section 3.4.1.4 showed a successful enrichment of *Sr*-Gal-1, -2 and -3. However it was not possible to remove some other proteins like the immunodiagnostic antigen L3NieAg, a metalloprotease, a petidyl-prolyl cis/trans isomerase and a troponin fragment. Among the mentioned proteins *Sr*-Gal-1 and -2 showed a high abundance according to their *unused protein scores* of 26.31 and 37.77 respectively. The results show that the galectins do bind β-galactoside structures but it was not possible to capture the previously unidentified galectins *Sr*-Gal-11, -21 and -22. In case the reason lies in the expression of these galectins at a very low level it should be possible to quantify the relative expression levels by using a comparative real time PCR method in future studies. Even though it seems attractive to isolate native galectins or galectin mixtures for immunological studies and assays it is still necessary to increase the amounts of worm material in order to obtain sufficient amounts of protein at the end.

A method that shows both the carbohydrate binding properties and the immunogenicity of *S. ratti* secreted galectins is the carbohydrate microarray method (Horlacher, 2008) presented in section 3.4.1.7. The test showed that galectins present in iL3 extracts are capable of binding to different carbohydrate structures. To some extend these results could apply for the E/S products due to the fact that Galectins were found in E/S products as well as in extracts. Since rat sera from infected rats were used to visualise the bound galectins it was shown that the host develops galectin specific antibodies during the course of the infection. In order to further study the subtle differences in carbohydrate-binding patterns it is necessary to further fractionate the native galectins or to express them singularly and test them as pure protein. This will lead to the identification of different endogenous ligands recognised by the various galectin types and, ultimately, their distinct biological roles. By doing so it has for example been shown that laminin is the most likely endogenous ligand for human galectin-1, a lectin proposed to mediate muscle development and induce apoptosis of activated T-cells by binding with CD45 glycoprotein (Perillo, 1995).

Since it was one aim of this work to recombinantly express an identified protein contained in the culture supernatants a member of the galectin family was chosen for expression in *E. coli*. First attempts to express galectin sequences *Sr*-Gal-1 or -2 were not successful due to unsatisfactory cloning results. *Sr*-Gal-3 was the first sequence that was successfully cloned and expressed as shown in section 3.4.1.5. The recombinant *Sr*-Gal-3 was tested against rat sera using ELISA. The results showed an immune recognition of the recombinant protein. Interestingly also sera from humans previously infected with *S. stercoralis* were capable of recognising the *S. ratti* galectin. This is probably related to the fact that the *Sr*-Gal-3 shows a 90% homology to the corresponding *S. stercoralis* EST cluster. Since the remaining galectins all show high homologies between 83–96% it is suitable to recombinantly express the remaining galectins in the future in order to study their immunogenic potential in both *S. ratti* and *S. stercoralis* infection.

The results show that *S. ratti* galectins are interesting candidates for further studies. Combining these findings with data on the different immunogenic capacities of nematode galectins that can be found in the literature they seem interesting for two reasons. Firstly they seem to be good candidates for vaccination strategies. For example Yanming *et al.* (2007) have shown that the application of recombinant *H. contortus* galectin let to a partial protection to homologous infection in goats. Secondly they might be good candidates to study inflammatory reactions since it has been shown that different members of the human galectin families provide inhibitory or stimulatory signals to control intestinal immune response under intestinal inflammatory conditions. Due to the fact that parasitic nematodes are also capable of masking or tuning the host immune response it is likely that galectins are involved in these processes.

4.4 The *S. ratti* POP

4.4.1 Role and classification of POPs

In vertebrates prolyl oligopeptidase (POP) activity has been found throughout the body, with a high concentration within the brain (Irazusta, 2002). This finding, along with the fact that several neuropeptides with internal proline residues, are recognised by POPs, led to a linkage of POP activity to a variety of neurological disorders such as Alzheimer's disease, amnesia, depression, and schizophrenia as well as other diseases such as blood pressure regulation, anorexia, bulimia nervosa, and Chagas' disease caused by *Trypanosoma cruzi* (Polgar, 2002). POPs have also been studied as a potential therapeutic agent for the treatment of celiac sprue, an inflammatory disease of the small intestine that is triggered by dietary, proline-rich gluten. POPs have previously not been associated with parasite infections.

The family of POPs was first described in 1991 based on the amino acid sequence homology of POP (E.C. 3.4.21.26), dipeptidyl peptidase IV (E.C. 3.4.14.5), acylaminoacyl peptidase (E.C. 3.4.19.1) (Rawlings, 1991) and oligopeptidase B (E.C. 3.4.21.83) (see figure 4.4.1-*1*). The POP group called family S9 has been grouped with other families into the serine carboxypeptidase (SC) clan of serine peptidases and are considered as members of the α/β-hydrolase enzymes. SC clan enzymes are endo- and exopeptidases that all share the catalytic triad composed of serine, aspartic acid and histidine. The cyclic amino acid proline plays a critical physiological

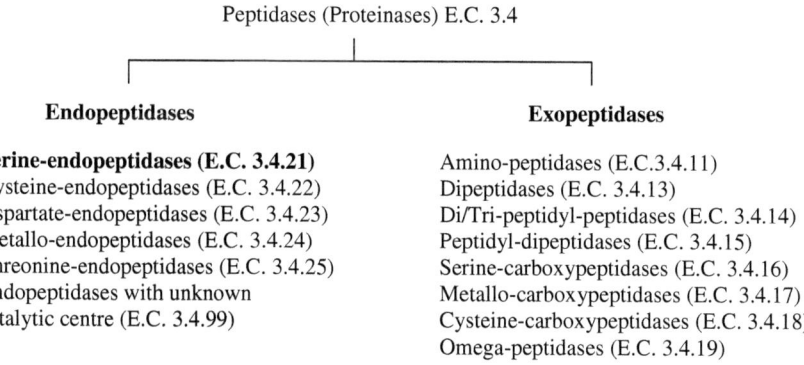

Figure 4.4.1-*1* Classification of peptidases/proteinases. POPs belong to the E.C. 3.4.21 group of serine proteases (Möhrlen, 2002)

role by protecting peptides from proteolytic degradation. The 80 kDa POPs are able to hydrolyse the peptide bond on the carboxyl side of internal proline residues. The catalytic triad within the catalytic domain consisting of serine, aspartic acid and histidine is essential for the activity of this enzyme.

4.4.2 *Sr*-POP identified in parasitic female E/S products and extracts

Evaluation of the sequence data from parasitic females showed that the culture supernatants and extracts contained sequences matching to the EST clusters SR01641 and SR03191. BLAST searches showed homologies with POPs. POPs were previously not reported in connection with parasitic helminth infections. Other non-nematode parasites that also express POPs are *T. brucei* and *L. major* (Venäläinen, 2004). Since it was one of the objectives to find novel stage-specific proteins that might have relevance for the containment of parasitism it was decided to further investigate this previously termed hypothetical protein of *S. ratti*. As presented in section 3.4.2 it was possible to identify a third cluster that aligned with the other two clusters and by performing RACE-PCR it was possible to capture the 3'- and 5'-ends of this secreted serine protease. BLAST searches showed that *B. malayi* is the only other parasitic nematode having a POP sequence in its genome. It is likely that this number will increase as soon as genomic information from other nematodes becomes available since POPs are present in most organisms and tissues (Polgar, 2002a). The sequence analysis led to the identification of the catalytic triad being necessary for the proteins' ability to cleave peptides consisting of not more than 30 amino acids after proline residues. In the microarray analysis of gender- and parasite-specific gene transcription in *S. ratti* Evans *et al.* did not report about any of the three clusters to be expressed predominantly in the parasitic females. This could be related to the possibility that the mentioned transcripts were not bound on the microarray. The detection of stage specificity in the presented work is supported by the fact that the ESTs can only be found in the EST libraries from parasitic females and that the comparative PCR presented in section 3.3.2 showed a positive result for parasitic female cDNA only; in iL3 cDNA no amplificate was obtained. As previously discussed the EST cluster SR04440 was found in a microarray analysis in samples from parasitic females under high immune pressure. The cluster is homologous to a POP, however, since it is not homologous to the newly identified sequence it is likely that the female express at least two different POP proteins. PCR experiments with this second peptidase were not successful which can be explained that the parasitic females used for all experiment were harvested at the sixth and seventh day post infection. The second POP cluster SR04440 however was found to be expressed at a higher level in females harvested 15 days post infection.

Investigation of previously published data showed that different POP inhibitors were synthesised and tested *in vitro* and also *in vivo* in both humans and rodents. The high occurrence of POP in the brain suggested that it is involved in the maturation and degradation of peptide hormones and neuropeptides, such as substance P, oxytocin, vasopressin and angiotensins, which are substrates for the enzyme. Studies in amnesia, depression and Alzheimer's disease have provided support for this hypothesis (Rosenblum, 2003). POP inhibitors have been shown to increase the brain levels of several neuropeptides, to reverse scopolamine-induced amnesia in rats and to improve cognition in old rats (Venäläinen, 2006). However, no studies with parasitic nematodes and POP inhibitors have been performed before. Thus it was decided to investigate which impact POP inhibitors might have on *in vitro* cultured parasitic females. The structures of the substances are shown in section 3.4.2.5; they all share the pyrrolidine structure and carry different substituents. Two additional substances were supplied but were not tested due to their insufficient solubility caused by a *tert*-butyl residue covalently bound to the pyrrolidine structure. Surprisingly all structures showed a lethal effect on the females as shown in figure 3.4.2.4.2. In samples treated with structures 2A and 2B the effect was already visible after 30 minutes at a concentration of 7.5 mM. *In vivo* studies have not been performed since the permit for animal studies with *S. ratti* did not cover the administration of active ingredients.

Nevertheless the *in vivo* studies showed a putative new mechanism of action against parasitic nematodes compare to known anthelmintics. *In vivo* studies will help to evaluate whether POP inhibitors will have the potential for a new substance class to treat infections with intestinal helminths. For latter tests it would be useful to apply the substances as sustained release formulations in order to ensure that the inhibitors are released at their location of action, the small intestine. In case the substances show an effect it might also be an option to treat infections with tissue dwelling nematodes since *B. malayi* also has a POP enzyme in its genome. It should however also be considered that these substances can also pass the blood brain barrier as shown in recent studies (Venäläinen, 2006) and thus might have cognitive side effects.

5 Abstract

Strongyloidiasis is one of the important intestinal nematode diseases and it is estimated that worldwide between 50-200 million people are infected by the human pathogen *S. stercoralis*. For research purposes the closely related parasitic nematode *S. ratti* can be used to study stage development, life cycle and parasitism. The high genetic similarity then allows to draw parallels between the human parasite and other Strongylid nematodes.

This work presents the establishment of the *S. ratti* life cycle at the BNI-Hamburg and the proteomic analysis of E/S and extract proteins from various developmental stages at the Proteomics Center, Boston. For the production of sufficient amounts of proteins protocols for the preparation of the developmental stages and the culture supernatants were implemented. Assuming that the protein patterns of larval and adult stages show differences, the results of the mass spectrometric analysis were compared. By choosing this approach it was possible to identify 1,081 proteins of *S. ratti*. In the EST analyses previously performed with *S. stercoralis* and *S. ratti* the proteins were previously termed hypothetical. Many of the proteins were also shown to be present as homologues in the culture supernatants and extracts of other parasitic nematodes and some proteins seem to be *Strongyloides*-specific. Under the scope to identify and describe excretory/secretory proteins from the parasitic nematode *S. ratti* which are abundantly or differentially produced by variable developmental stages in the life cycle, distinct proteins/protein family were chosen for further investigation.

It was possible to show that members of the galectin family are highly abundant and secreted by various stages. Screening and comparing the protein and EST databases led to the identification of seven different galectin sequences. Four of these proteins were sequenced completely and two were partially sequenced. It was possible to recombinantly express one galectin family member in *E. coli* and ELISA analysis showed antibody recognition by *S. ratti* and *S. stercoralis* immune sera. Carbohydrate microarrays showed the binding of galectins to different carbohydrate structures.

The second protein that was chosen is *Sr*-POP-1. This newly discovered serine protease is secreted stage-specific by parasitic females only. Screening and comparison of EST sequences with subsequent RACE PCR analysis led to the identification of the whole 91 kDa protein. 3-D modelling revealed a comparative structure with other reported POPs and showed the putative mode of enzymatic action. Previous reports about effective inhibition of human and rodent POP enzymes with specific agents led to the assumption that these agents might also have an impact on female behaviour when cultured *in vitro*. The addition of the inhibitors led to the effective killing of parasitic females.

The presented work shows the value of the proteomic approach in order to identify new proteins in parasites that might offer new perspectives for the treatment of parasitic infections or for the development of vaccine candidates. In addition it is possible to identify candidates that contribute to the containment of parasitism and influence host immune responses.

6 Zusammenfassung

Strongyloidiasis ist eine der bedeutendsten Infektionen mit intestinalen parasitären Nematoden. Schätzungen zufolge liegt die Zahl der weltweit mit dem humanen Parasiten *Strongyloides stercoralis* infizierten Individuen zwischen 50 bis 200 Millionen. Zu derselben Gattung gehört der nahe verwandte Parasit von Ratten *Strongyloides ratti*, der sehr gut geeignet ist, die Stadienentwicklung, den Lebenszyklus und für die parasitische Lebensweise relevanten Moleküle im Labor zu studieren. Durch die genetische Verwandschaft zu *S. stercoralis* ist es möglich, aus Ergebnissen, die mit *S. ratti* gewonnen werden, Rückschlüsse auf den Humanparasiten und andere Nematoden der Gattung *Strongyloides* zu ziehen.

In der vorliegenden Arbeit wird die Etablierung des Modellparasiten *S. ratti* am Bernhard-Nocht-Institut in Hamburg bis zu der Proteomanalyse von exkretorisch/sekretorischen (E/S)- und Extraktproteinen verschiedener Entwicklungsstadien am Proteomics Center (Children's Hospital, Boston, USA) dargestellt. Um ausreichende Proteinmengen zu gewinnen, wurden zunächst die notwendigen Methoden für die Kultivierung verschiedener parasitär lebender und nicht parasitärer Entwicklungsstadien erarbeitet. Durch die anschließenden massenspektrometrischen Analysen konnten insgesamt 1.086 Proteine identifiziert werden. Der Vergleich der einzelnen Stadien hat gezeigt, dass sich deren Proteinmuster unterscheiden. Zugleich konnte nachgewiesen werden, dass die bisher nur als „Expressed Sequence Tags" (ESTs) identifizierten Sequenzen zu Proteinen translatiert werden. In vergleichbaren aus der Literatur bekannten Analysen wurden einerseits einige der E/S-Proteine bei anderen parasitären Nematoden ebenfalls als Homologe identifiziert, andererseits konnten *Strongyloides*-spezifische Proteine identifiziert werden. Da es das Ziel war, abundante oder differentiell exprimierte Proteine zu identifizieren und zu beschreiben, wurden im Anschluss Proteine der Galektin-Familie und eine Prolyl-oligopeptidase (POP) molekularbiologisch untersucht.

Es wurde gezeigt, dass Mitglieder der Galektin-Familie abundant sekretierte Proteine verschiedener Entwicklungsstadien darstellen. Mit Hilfe der verfügbaren EST-Datenbanken konnten sieben verschiedene *S. ratti*-Galektinsequenzen identifiziert werden, von denen vier vollständig und zwei teilweise sequenziert werden konnten. Ein Mitglied der Galektin-Familie wurde in *Escherichia coli* rekombinant exprimiert und im ELISA wurde dessen Erkennung durch Antikörper in Seren von *S. ratti*-infizierten Ratten bzw. *S. stercoralis*-infizierten Menschen nachgewiesen. Die Bindung von in den E/S-Produkten enthaltenen Galektinen an verschiedene β-galaktosidische Strukturen wurde mit Hilfe von Kohlenhydrat-Mikroarrays bewiesen.

Als stadienspezifisch exprimiertes Protein wurde das Enzym Prolyl-oligopeptidase (*Sr*-POP) identifiziert und beschrieben. Diese in *S. ratti* neu entdeckte Serinprotease wird ausschließlich von parasitären Weibchen sekretiert. Ein Vergleich von EST-Daten mit anschließender „Rapid Amplification of cDNA Ends" (RACE)-PCR-Analyse hat zur vollständigen Sequenz des Enzyms von 91 kDa geführt. Das 3-D-Modelling des *S. ratti* Proteins bestätigte die charakteristische POP-Struktur, die vereinbar ist mit dem beschriebenen POP-Reaktionsmechanismus. Da publiziert wurde, dass POPs anderer Spezies mit spezifischen Inhibitoren gehemmt werden können, wurden an parasitären *S. ratti*-Weibchen Inhibitionsversuche mit verschiedenen Hemmern durchgeführt, die für eine eine anti helminthische Wirkung *in vitro* sprechen.

Die Ergebnisse der vorliegenden Arbeit zeigen, daß durch die Anwendung massenspektrometrischer Methoden Kandidatenproteine identifiziert werden können, die z.B. für parasitäre Stadien charakteristisch und von funktioneller Bedeutung sein können und möglicherweise die Immunantwort des Wirtes beeinflussen können. Diese Kandidatenproteine können in der Zukunft neue Ansätze für die Behandlung parasitärer Erkrankungen oder für die Entwicklung von Vakzinen bieten.

7 Acknowledgements

Foremost, I want to express my special gratitude to my supervisor PD Dr. Norbert W. Brattig who gave me the opportunity to become a PhD student at his department and to work with this challenging nematode parasite. He has been guiding the work over the entire period of three years and supported me with his confidence and ideas whenever possible.

Furthermore, I am grateful to Dr. Hanno Steen and his wife, Judith Jebanathirajah Steen, for the great opportunity to work at the Proteomics Center at Children's Hospital Boston, MA, USA for several months. Also I would like to thank Hanno for his suggestions regarding the manuscript and his endurance in reminding me to finish the work.

Also I would like to thank Prof. Peter Heisig for taking interest in this thesis and for accepting to review it.

For the help with parts of my work I particularly would like to thank Yvette Endriss from the STI, Basel, for her help to get the cycle started, Makedonka Mitreva, PhD, from the Genome Sequencing Center, St. Louis, for the *Strongyloides* databases, Elina Jarho from the University of Kuopio, for providing the POP inhibitors and Tim Horlacher from the Seeberger Glycomic research group, Zürich, for performing the carbohydrate arrays.

Thanks to the Brattig laboratory: Frank Geisinger (thanks for the B. Spears newsletters), Silke van Hoorn, Kerstin Krausz and Hassan Mohammed for the lab support and Nadine Borchert for introducing me to the secrets of molecular biology. Also thanks to Astrid Mewes, Fabian Imse and Thomas Schulze for their help with the cycle and the support in the laboratory. At the Proteomics Center I would like to thank Yin-Yin Lin and Zachary Waldon for help and instructions regarding the mass spectrometry work.

I am also thankful to Yasmina Tazir, Sven Liffers and Flavio Monigatti for becoming good friends during the last three years and Sven and Flavio also for the nice accommodation in Boston.

I am grateful to the Vereinigung der Freunde des Tropeninstituts Hamburg e.V. and the Boehringer Ingelheim Fonds for the partial scholarships.

Thanks also to Wilhelm Wollschläger who sparked my interest in science and whose slogan "viel hilft viel" proved to be true in most of the cases.

For proof-reading the thesis I want to thank my sister, Sabine Akerman, and my whole family for their everlasting support.

Finally, I want to thank my wife, Sabine, for the last ten years and hopefully for many more to come.

8 References

Albonico, M., Cropmton, D.W., Savioli, L. 1999: Control strategies for human intestinal nematode infections. Advances in Parasitology, 42, 277-341.

Anthony, R.M., Rutitzky, L.I., Urban Jr, J.F., Stadecker, M.J., Gause, W.C. 2007: Protective immune mechanisms in helminth infection. Nature reviews, 7, 975-987.

Arnold, K., Bordoli, L., Kopp, J., Schwede, T. 2006: The SWISS-MODEL Workspace: A web-based environment for protein structure homology modelling. Bioinformatics, 22,195-201.

Ashford, R.W., Barnish, G. 1989: *Strongyloides fuelleborni* and similar parasites in animals and man. In: Grove, D.I. (ed): Strongyloidiasis: a major roundworm infection in man, 271-286, Taylor and Francis, London.

Atack, J.R., Suman-Chauhan, N., Dawson, G., Kulagowski, J.J. 1991: *In vitro* and *in vivo* inhibition of prolyl endopeptidase. European Journal of Pharmacology, 205, 157-163.

Awasthi, S., Bundi, D.A.P., Savioli, L. 2003: Helmintic infections. British Medical Journal, 327, 431-433.

Barondes, S.H., Castronovo, V., Cooper, D.N., Cummings, R.D., Drickamer, K., Feizi, T.,Gitt, M.A., Hirabayashi, J., Hughes, C., Kasai, K., et al. 1994: Galectins: A family of animal beta-galactoside-binding lectins. Cell, 76, 597-598.

Basavaraju, S., Zhan, B., Kennedy, M.W., Liu, Y., Hawdon, J., Hotez, P.J. 2003: Ac-FAR-1, a 20 kDa fatty acid- and retinol-binding protein secreted by adult *Ancylostoma caninum* hookworms: gene transcription pattern, ligand binding properties and structural characterisation. Molecular and Biochemical Parasitology, 126, 63-71.

Beaudoin, A.R., Grondin, G. 1991: Shedding of vesicular material from the cell surface of eukaryotic cells: different cellular phenomena. Biochimica et Biophysica Acta, 1071, 203-219.

Blaxter, M.L., De Ley, P., Garey, J.R., Liu, L.X.,Scheldeman, P., Vierstraete, A. Vanfleteren, J.R., Mackey, L.Y., Dorris, M., Frisse, L.M., Vida, J.T., Kelley Thomas, W. 1998: A molecular evolutionary framework for the phylum nematoda. Nature, 392, 71-75.

Bleay, C., Wilkes, C.P., Paterson, S., Viney, M. 2007: Density-dependent immune responses against the gastrointestinal nematode *Strongyloides ratti*. International Journal for Parasitology, 37, 1501-1509.

Borchert, N. 2006: Identifizierung und Charakterisierung von gewebsspaltenden Proteinsaen bei parasitären Nematoden. PhD thesis.

Borchert, N., Becker-Pauly, C., Wagner, A., Fischer, P., Stöcker, W., Brattig, N.W. 2007: Identification and characterization of onchoastacin, an astacin-like metalloproteinase from the filaria *Onchocerca volvulus*. Microbes and Infection, 9, 498-506.

Buechner, M., Hall, D.H., Bhatt, H., Hedgecock, E.M. 1999: Cystic canal mutants in *Caenorhabditis elegans* are defective in the apical membrane domain of the renal (excretory) cell. Developmental Biology, 214, 227-241.

Compton, D.W.T. 1987: Human helmintic populations. In: Pawlowski, Z.S. (ed): Bailliere's clinical tropical medicine and communicable diseases, vol.2, 489-510, Bailliere Tindall, London.

Craig, H., Wastling, J.M., Knox, D.P. 2006: A preliminary proteomic survey of the *in vitro* excretory/secretory products of fourth-stage larval and adult *Teladorsagia circumcincta*. Parasitology, 132, 535-543.

Dereeper, A., Guignon, V., Blanc, G., Audic, S., Buffet, S., Chevenet, F., Dufayard, J.F., Guindon, S., Lefort, V., Lescot, M., Claverie, J.M., Gascuel, O. 2008: Phylogeny.fr: robust phylogenetic analysis for the non-specialist. Nucleic Acids Research, 36, W465-W469.

Evans, H., Mello, L.V., Fang, Y., Wit, E., Thompson, F.J., Viney, M.E. & Paterson, S. 2008: Microarray analysis of gender- and parasite-specific gene transcription in *Strongyloides ratti*. International Journal for Parasitology, 38, 1329-1341.

Genta, R.M., Caymmi Gomes, M. 1989: Pathology. In: Grove, D.I. (ed): Strongyloidiasis: a major roundworm infection in man, 105-132, Taylor and Francis, London.

Gomez Gallego, S., Loukas, A., Slade, R.W., Neva, F.A., Varatharajalu, R., Nutman, T.B., Brindley, P.J. 2005: Identification of an astacin-like metallo-proteinase transcript from the infective larvae of *Strongyloides stercoralis*. Parasitology International, 54, 123-133.

Greenbaum, D., Luscombe, N.M., Jansen, R., Qian, J., Gerstein, M. 2001: Interrelating different types of genomic data, from proteome to secretome: 'oming in on function. Genome Research, 11, 1463-1468.

Greenhalgh, C.J., Beckham, S.A., Newton, S.E. 2000: Galectins from sheep gastrointestinal nematode parasites are highly conserved. Molecular and Biochemical Parasitology, 98, 285-289.

Harder, A. 2002: Chemotherapeutic approaches to nematodes: current knowledge and outlook. Parasitology Research, 88, 272-277.

Hartmann, S., Sereda, M.J., Sollwedel, A., Kallina, B., Lucius, R. 2006: A nematode allergen elicits protection against challenge infection under specific conditions. Vacine, 24, 3581-3590.

Harvey, S.C., Gemmill, A.W., Read, A.F., Viney, M.E. 2000: The control of morph development in the parasitic nematode *Strongyloides ratti*. Proceedings of the Royal Society, 267, 2057-2063.

Harvey, S.C., Viney, M.E. 2001: Sex determination in the parasitic nematode *Strongyloides ratti*. Genetics, 158, 1527-1533.

Hauber, H.P., Galle, J., Chiodini, P.L., Rupp, J., Birke, R., Vollmer, E., Zabel, P., Lange, C. 2005: Fatal Outcome of a Hyperinfection Syndrome despite Successful Eradication of *Strongyloides* with Subcutaneous Ivermectin. Infection, 33, 383-386.

Hewitson, J.P., Harcus, Y.M., Curwen, R.S., Dowle, A.A., Atmadja, A.K., Ashton, P.D., Wilson, A., Maizels, R.M. 2008: The secretome of the filarial parasite, *Brugia malayi*: Proteomic profile of adult excretory-secretory products. Molecular and Biochemical Parasitology, 160, 8-21.

Horlacher, T., Seeberger, P.H. 2008: Carbohydrate arrays as tools for research and diagnostics. Chemical Society Reviews, 37, 1414–1422.

Hughes, R.C. 1999: Secretion of the galectin family of mammalian carbohydrate-binding proteins. Biochimica et Biophysica Acta, 1473, 172-185.

References

Irazusta, J., Larrinaga, G., Gonzales-Maesto, J., Gil, J., Meana, J.J., Casis, L. 2002: Distribution of prolyl endopeptidase activities in rat and human brain. Neurochemistry International, 40, 337-345.

Jenkins, R.E., Taylor, M.J., Gilvary, N.J., Bianco, A.E. 1998: Tropomyosin implicated in host protective responses to microfilariae in onchocerciasis. Immunology, 95, 7550-7555.

Jiang, D., Li, B.W., Fischer, P.U., Weil, G.J. 2008: Localization of gender-regulated gene expression in the filarial nematode *Brugia malayi*. International Journal of Parasitology, 38, 503-512.

Johnston, M.J.G., MacDonald, J.A., McKay, D.M. 2009: Parasitic helminths: a pharmacopeia of anti-inflammatory molecules. Parasitology, 136, 125-147.

Kaplan, R.M. 2004: Drug resistance in nematodes of veterinary importance: a status report. Trends in Parasitology, 20, 477-481.

Klion, A.D., Donelson, J.E. 1994: *Ov*GalBP, afilarial antigen with homology to vertebrate galactoside-binding proteins. Molecular and Biochemical Parasitology, 65, 305-315.

Klionsky, D.J. 1998: Nonclassical protein sorting to the yeast vacuole. Journal of Biological Chemistry, 273, 10807-10810.

Knudsen, G.M., Medzihradszky, K.F., Lim, K.C., Hansell, E., McKerrow, J.H. 2005: Proteomic Analysis of *Schistosoma mansoni* Cercarial Secretions. Molecular and Cellular Proteomics, 4, 1862-1875.

Ko, R.C., Fan, L. 1996: Heat shock response of *Trichinella spiralis* and *T. pseudospiralis*. Parasitology, 112, 89-95.

Laemmli, U.K. 1970: Cleavage of structural proteins during the assembly of the head of bacteriophage T4. Nature, 227, 680-685.

Leffler, H., Carlsson, S., Hedlund, M., Qian, Y., Poirer, F. 2004: Introduction to Galectins. Glycoconjugate Journal, 19, 433-440.

Li, M., Chen, Z., Lin, X., Zhang, X., Song, Y., Wen, Y., Li, J. 2008: Engineering of avermectin biosynthetic genes to improve production of ivermectin in Streptomyces avermitilis. Bioorganic & Medicinal Chemistry Letters, 18(20), 5359-5363.

Lightowlers, M.W., Rickard, M.D. 1988: Excretory-secretory products of helminth parasites: effects on host immune responses. Parasitology, 96, S123-S166.

Lim, S., Katz, K., Krajden, S., Fuksa, M., Keystone, J.S., Kain, K.C. 2004: Complicated and fatal *Strongyloides* infection in Canadians: risk factors, diagnosis and management. Canadian Medical Association Journal, 171, 479-484.

Lok, J.B. 2007: *Strongyloides stercoralis*: a model for translational research on parasite nematode biology. WormBook, http://www.wormbook.org.

Maizels, R.M., Bundy, D.A.P., Selkirk, M.E., Smith, D.F., Anderson, R.M. 1993: Immunological modulation and evasion by helminth parasites in human populations. Nature, 365, 797-805.

Maizels, R.M., Yazdanbakhsh, M. 2003: Immune regulation by helminth parasites: cellular and molecular mechanisms. Nature Reviews Immunology, 3, 733-744.

McKerrow, J.H., Pino-Heiss, S., Lindquist, R.L., Werb, Z. 1985: Purification and characterization of an elastinolytic proteinase seceted by cercariae of *Schistosoma mansoni*. Journal of Biological Chemistry, 260, 3703-3707.

Mimori, T., Tanaka, M., Tada, I. 1987: *Strongyloides ratti*: Formation of Protection in Rats by Excretory/Secretory Products af Adult Worms. Experimental Parasitology, 64, 342-346.

Mir, A., Benahmed, D., Igual, R., Borras, R., O´Connor, J.E., Moreno, M.J., Rull, S. 2006: Eosinophil-selective Mediators in Human Strongyloidiasis. Parasite Immunology, 28, 397-400.

Möhrlen, F. 2002: Analyse der Struktur, Funktion und Evolution der Astacin-Protein-Familie. PhD thesis.

Nagaraj, S.H., Gasser, R.B., Ranganathan, S. 2008: Needles in the EST Haystack: Large-Scale Identification and Analysis of Excretory-Secretory (ES) Proteins in Parasitic Nematodes Using Expressed Sequence Tags (ESTs) PLOS Neglected Tropical Diseases, 2, e301.

Nolan, T.J., Bhopale, V.M., Schad, G.A. 1999: Hyperinfective strongyloidiasis: *Strongyloides stercoralis* undergoes an autoinfective burst in neonatal gerbils. Journal of Parasitology, 85, 286/289.

World Health Organization 2003: Action against worms (issue 1).

Pelletier, I., Hashidate, T., Urashima, T., Nishi, N., Nakamura, T., Futai, M., Arata, Y., Kasai, K., Hirashima, M., Hirabayashi, J., Sato, S. 2003: Specific recognition of *Leishmania major* poly-beta-galactosyl epitopes by galectin-9: possible implication of galectin-9 in interaction between *L. major* and host cells. Journal of Biological Chemistry, 278, 22223-22230.

Perillo, N.L., Pace, K.E., Seilhamer, J.J., Baum, L.G. 1995: Apoptosis of T cells mediated by galectin-1. Nature, 378, 736-739.

Polgar, L. 2002: Structure-function of prolyl oligopeptidase and its role in neurological disorders. Current Medicinal Chemistry - Central Nervous System Agents, 2, 251-257.

Polgar, L. 2002a: The prolyl oligopeptidase family. Cellular and Molecular Life Sciences, 59, 349-362.

Porto, A.F., Neva, F.A., Bittencourt, H., Lisboa, W., Thompson, R., Alcântara, L., Carvalho, E.M. 2001: HTLV-1 decreases Th$_2$ type of immune response in patients with strongyloidiasis. Parasite Immunology, 23, 503-507.

Pou-Barreto, C., Quispe-Ricalde, M.A., Morchon, R., Vazquez, C., Genchi, M., Postigo, I., Valladares, B., Simon, F. 2008: Galectin and aldolase-like molecules are responsible for the specific IgE response in humans exposed to *Dirofilaria immitis*. Parasite Immunology, 30, 596-602.

Ramanathan, R., Burbelo, P.D., Groot, S., Iadarola, M.J., Neva, F.A., Nutman, T.B. 2008: A luciferase immunoprecipitation systems assay enhances the sensitivity and specificity of diagnosis of *Strongyloides stercoralis* infection. Journal of Infectious Dieseases, 198, 444-451.

Rawlings, N.D., Polgár, L., Barret, A.J. 1991: A new family of serine-type peptidases related to prolyl oligopeptidase. Biochemical Journal, 279, 907-908.

Rehm, H., Letzel, T. 2006: Elektrospray Ionisierung. In: Rehm, H., Der Experimentator - Proteinbiochemie / Proteomics, 243-247, Elsevier GmbH, München.

Rosenblum, J.S., Kozarich, J.W. 2003: Prolyl peptidases: a serine protease subfamily with high potential for drug discovery. Current Opinion in Chemical Biology, 7, 496-504.

Sambrook, J., Fritsch, E.F., Maniatis, T. 1989: Molecular Cloning - A Laboratory Manual, 2nd Edition, Cold Spring Harbor Laboratory Press.

Schad, G.A. 1989: Morphology and life history of *Strongyloides stercoralis*. In: Grove, D.I. (ed): Strongyloidiasis: a major roundworm infection in man, 85-104, Taylor and Francis, London.

Schatz, G., Dobberstein, B. 1996: Common principles of protein translocation across membranes. Science, 271, 1519-1526.

Schneider, J.S., Giardiniere, M., Morain, P. 2002: Effects of the prolyl endopeptidase inhibitor S 17092 on cognitive deficits in chronic low dose MPTP-treated monkeys. Neuropsychopharmacology, 26, 176-182.

Schwarzbauer, J.E., Musset-Bilal, F., Ryan, C.S. 1994: Extracellular calcium-binding protein SPARC/osteonectin in *Caenorhabditis elegans*. Methods in Enzymology, 245, 257-270.

Steinmann, P., Zhou, X., Du, Z., Jiang, J., Xiao, S., Wu, Z., Zhou, H., Utzinger, J. 2008: Tribendimidine and Albendazole for treating soil transmitted helminths, *Strongyloides stercoralis* and *Taenia* spp.: open-label ranodomized trial. PLOS Neglected Tropical Diseases, 2, 1-10.

Stephenson, L.S., Latham, M.C., Ottesen, E.A. 2000: Malnutrition and parasitic helminth infections. Parasitology, 121, S23-S38.

Szeltner, Z., Polgar, L. 2008: Structure, Function and Biological Relevance of Prolyl Oligopeptidase. Current Protein and Peptide Science, 9, 96-107.

Taub, L.M., Kornfeld, S. 1997: The trans-Golgi network: a late sorting station. Current Opinion in Cell Biology, 9, 527-533.

Terada, M., Sano, M. 1985: Effects of diethylcarbamazine on the motility of *Angiostrongylus cantonensis* and *Dirofilaria immitis*. Parasitology Research, 72, 375-385.

Thompson, F.J., Barker, G.A., Hughes, L., Viney, M.E. 2008: Genes important in the parasitic life of the nematode *Strongyloides ratti*. Molecular and Biochemical Parasitology, 158, 112-119.

Thompson, F.J., Barker, G.L.A., Hughes, L., Wilkes, C.P., Coghill, J., Viney, M.E. 2006: A microarray analysis of gene expression in the free-living stages of the parasitic nematode *Strongyloides ratti*. BMC Genomics, 7, 157.

Thompson, F.J., Cockroft, A.C., Wheatley, I., Britton, C., Devaney, E. 2001: Heat shock and developmental expression of hsp83 in the filarial nematode *Brugia pahangi*. European Journal of Biochemistry, 268, 5808-5825.

Thompson, F.J., Mitreva, M., Barker, G.L.A., Martin, J., Waterson, R.H., McCarter, J.P., Viney, M.E. 2005: An expressed sequence tag analysis of the life cycle of the parasitic nematode *Strongyloides ratti*. Molecular and Biochemical Parasitology, 142, 32-46.

Thomson, D.P., Geary, T.G. 2002: Excretion/secretion, ionic and osmotic regulation. In: Lee, D.L., The Biology of Nematodes, 291-320, Taylor and Francis, London.

Toide K., Shinoda, M., Fujiwara, T., Iwamoto, Y. 1997: Effect of a novel prolyl endopeptidase inhibitor, JTP-4819, on spatial memory and central cholinergic neurons in aged rats. Pharmacology Biochemistry and Behavior, 56, 427-434.

Toide, K., Fujiwara, T., Iwamoto, Y., Shinoda, M., Okamiyz, K., Kato, T. 1996: Effect of a novel prolyl oligopeptidase inhibitor, JTP-4819, on neuropeptide metabolism in the rat brain. Naunyn-Schmiedeberg's Archives of Pharmacology, 353, 355-362.

Toscano, M.A., Commodaro, A.G., Ilarregui, J.M., Bianco, G.A., Libermann, A., Serra, H.M., Hirabayashi, J., Rizzo, L.V., Rabinovich, G.A. 2006: Galectin-1 suppresses autoimmune retinal disease by promoting concomitant Th_2- and T regulatory-mediated anti-inflammatory responses. Journal of Immunology, 176, 6323-6332.

Turner, D.G., Wildblood, L.A., Inglis, N.F., Jones, D.G. 2008: Characterization of a galectin-like activity from the parasitic nematode, *Haemonchus contortus*, which modulates ovine eosinophil magration *in vitro*. Veterinary Immunology and Immunpathology, 122, 138-145.

Venäläinen, J.I., Garcia-Horsman, J.A., Forsberg, M.M., Jalkanen, A., Walle, E.A.A., Jarho, E.M., Christiaans, J.A.M., Gynther, J., Männisto, P.T. 2006: Binding kinetics and duration of in vivo action of novel prolyl oligopeptidase inhibitors. Biochemical Pharmacology, 71, 683-692.

Venäläinen, J.I., Juvonen, R.O., Männistö, P.T. 2004: Evolutionary relationships of the prolyl oligopeptidase family enzymes. European Journal of Biochemistry, 271, 2705-2715.

Viney, M.E. 1994: A genetic analysis of reproduction in *Strongyloides ratti*. Parasitology, 109, 511-515.

Viney, M.E., Brown, M., Omoding, N.E., Bailey, W., Gardner, M.P., Roberts, E., Morgan, D., Elliott, A.M., Whitworth, J.A.G. 2004: Why Does HIV Infection Not Lead to Disseminated Strongyloidiasis. Journal of Infectious Diseases, 190, 2175-2180.

Whitehead, A.G., Hemming, J.R. 1965: A comparison of some quantitative methods of extracting small vermiform nematodes from soil. Annals of Applied Biology, 55, 25-38.

Wilkes, C.P., Bleay, C., Paterson, S., Viney, M.E. 2007: The immune response during a *Strongyloides ratti* infection of rats. Parasite Immunology, 29, 339-346.

Wu, Z., Nagano, I., Boonmars, T., Takahashi, Y. 2007: Thermally induced and developmentally regulated expression of a small heat shock protein in *Trichinella spiralis*. Parasitology Research, 101, 201-212.

Yanming, S., Ruofeng, Y., Muleke, C.I., Gaungwei, Z., Lixin, X., Xiangrui, L. 2007: Vaccination of goats with recombinant galectin antigen induces partial protection against *Haemonchus contortus* infection. Parasite Immunology, 29, 319-326.

Yazdanbakhsh, M., Kremsner, P.G., van Ree, R. 2002: Allergy, parasites, and the hygiene hypothesis. Science, 296, 490-494.

Young, A.R., Barcham, G.J., Kemp, J.M., Dunphy, J.L., Nash, A., Meeusen, E.N. 2008: Functional Characterization of an eosinophil-specific galectin, ovine galectin-14. Glycoconjugate Journal, 26, 423-432.

9 Appendices

9.1 Protein Lists

9.1.1 Table 1a: List of *Strongyloides* EST cluster numbers found in E/S products from the parasitic, the infective and the free-living stages

	Cluster	BLAST Alignment	Species	Accession Number	E	SP	EST Lgt.	% Cov.	# Pep.	UPS
Oxydative metabolism										
1	SS01309*	Peroxiredoxin	*A. suum*	Q9NL98	$1e^{-70}$	no	196	49.1	4	12.45
2	SS01223	Superoxide dismutase	*C. elegans*	NP_492290	$8e^{-69}$	no	173	52.5	7	14.00
3	SS01344*	Superoxide dismutase	*B. pahangi*	P41962	$1e^{-58}$	no	157	34.1	5	11.70
4	SR04901	Thioredoxin	*C. brenneri*	ACD86930	$7e^{-31}$	trun	129	46.0	6	12.01
5	SR04239*	Thioredoxin	*A. suum*	AAS78778	$1e^{-48}$	no	147	32.7	4	8.55
6	SR00374	Thioredoxin peroxidase 1	*B. malayi*	P48822	$2e^{-73}$	yes	179	17.2	2	5.79
7	SR03061	Glutathione peroxidase	*B. malayi*	XP_00189 75 17	$1e^{-39}$	no	144	33.9	4	8.00
8	SR04471	FAD dependent oxidoreductase	*B. malayi*	XP_00189 24 48	$3e^{-47}$	yes	195	25.6	4	8.05
Carbohydrate Metabolism										
9	SR00792	Probable citrate synthase	*C. elegans*	P34575	$1e^{-166}$	no	339	46.0	11	24.50
10	SR05207	Transaldolase	*N. vitripennis*	XP_00160 21 66	$2e^{-72}$	no	193	45.9	6	14.59
11	SR05258	Citrate synthase	*C. elegans*	NP_499264	$2e^{-88}$	no	196	40.3	4	4.01
12	SR03532	Triose phosphate isomerase	*C. elegans*	1MO0_A	$2e^{-45}$	no	129	34.3	3	4.00
13	SR02487	Carbohydrate phosphorylase	*B. malayi*	XP_00180 46 69	$3e^{-81}$	no	171	49.5	7	7.44
14	SR00093	UDP-glucose pyrophosphorylase	*B. malayi*	XP_00189 97 21	$1e^{-80}$	no	182	31.3	3	9.44
15	SR01173*	Fructose biphosphate aldolase 1	*C. elegans*	P54216	$1e^{-12}$	no	69	71.0	4	16.03
16	SR01890	Aldolase	*G. rostochiensis*	AAN78210	$2e^{-54}$	no	185	15.1	3	6.12
17	SR02615	Transketolase	*B. malayi*	XP_00189 35 28	$8e^{-75}$	no	193	51.3	9	16.00
18	SR03215	Transaldolase	*C. elegans*	NP_741369	$1e^{-68}$	no	178	36.5	5	12.76
19	SR01273	Aldehyde dehydrogenase (alh-12)	*C. elegans*	NP_00102 29 31	$1e^{-36}$	no	171	13.5	2	5.78
20	SS00861	Aldehyde dehydrogenase (alh-8)	*C. elegans*	NP_00102 20 78	$4e^{-123}$	no	272	22.4	4	9.54
21	SR00739	Aldehyde dehydrogenase (alh-8)	*C. elegans*	NP_00102 20 78	$2e^{-124}$	no	274	15.7	3	8.26
22	SS00910	GDP-L-fucose synthetase	*B. malayi*	XP_00189 37 25	$1e^{-62}$	no	154	42.9	3	6.92
23	SR01995	Aldo/keto reductase family protein	*B. malayi*	XP_00189 77 43	$2e^{-35}$	no	159	43.4	5	16.21

Table 1a continued

Cluster	BLAST Alignment	Species	Accession Number	E	SP	EST Lgt.	% Cov.	# Pep.	UPS
Cytosol energy metabolism									
24 SR00185*	Enolase (enol-1)	C. elegans	NP_001022349	$2e^{-77}$	no	184	40.8	5	10.05
25 SR00704*	Cytochrome C family member	C. elegans	NP_500629	$1e^{-46}$	no	106	45.2	4	8.09
26 SR02190*	Enolase (enol-1)	C. elegans	NP_001022349	$2e^{-78}$	no	167	57.5	5	8.35
27 SR01215	Fumarase	B. malayi	XP_001900957	$2e^{-70}$	no	168	39.9	4	10.16
28 SR00949	ATP synthase subunit	C. elegans	NP_498111	0.0	no	473	34.5	10	21.70
29 SR00265	Lyase	A. suum	AAP51177	$2e^{-117}$	no	241	49.4	6	13.42
30 SR00941	Aconitase family member	C. elegans	NP_498738	$1e^{-123}$	no	280	42.1	8	18.18
31 SR02997	Isocitrate dehydrogenase	T. castaneum	XP_968850	$1e^{-84}$	no	179	23.5	3	6.00
32 SR00966	Arginine kinase	H. glycines	AAO49799	$2e^{-165}$	no	346	61.8	19	41.48
Protein biosynthesis									
33 SR00985	Elongation factor family member	C. elegans	NP_492457	$1e^{-162}$	no	321	28.9	7	15.60
34 SR00885	Eukaryotic translation initiation factor 5A-2	B. malayi	XP_001902971	$8e^{-68}$	no	155	17.4	3	6.41
35 SR03540	TPR domain containing protein	B. malayi	XP_001892919	$3e^{-30}$	no	251	33.5	7	15.15
36 SR00180	Cyclophylin	C. elegans	NP_497297	$4e^{-51}$	yes	152	25.7	4	11.93
Protein digestion and folding									
37 SR00990*	Peptidyl-prolyl cis-trans isomerase	D. immitis	Q23955	$1e^{-59}$	no	162	56.2	8	17.14
38 SR01407	Protein disulfide isomerase	T. circumcinata	Q2HZY3	0.0	yes	287	35.9	10	23.95
39 SR00881	W07G4.4 Peptidase M17	C. elegans	NP_506260	$1e^{-111}$	no	356	31.5	11	25.77
40 SR01037*	Protein disulfide isomerase	A. suum	CAK18211	0.0	yes	499	62.7	27	62.56
41 SS00975	Protein disulfide isomerase (pdi-3)	C. elegans	NP_491995	$5e^{-63}$	yes	152	27.6	3	6.00
42 SR02132*	Serpin	B. malayi	XP_001896649	$1e^{-17}$	yes	162	45.1	7	15.36
43 SR02845	Serpin	B. malayi	XP_001892287	$2e^{-26}$	yes	173	28.3	4	9.87
44 SR02150	Serpin	M. musculus	AAR89288	$5e^{-24}$	no	184	51.9	7	14.00
45 SR01375*	Cyclophylin-type peptidyl-prolyl cis-trans isomerase 15	B. malayi	XP_001896264	$8e^{-70}$	no	186	49.5	8	16.74
46 SR01148*	FKBP-peptidyl-prolyl cis-trans isomerase	B. malayi	XP_001901266	$7e^{-39}$	no	143	58.0	7	17.23
47 SR01041	Calreticulin	N. americanus	CAA07254	$1e^{-153}$	yes	410	34.6	8	17.37
48 SR02163	Peptidase family M1 containing protein	B. malayi	XP_001897028	$4e^{-36}$	no	178	51.7	8	18.51

Table 1 a continued

	Cluster	BLAST Alignment	Species	Accession Number	E	SP	EST Lgt.	% Cov.	# Pep.	UPS
49	SR00921	Serine carboxy-peptidase	B. malayi	XP_00190088	9e^{-101}	yes	303	13.5	2	5.71
50	SR01946	Proteasome subunit beta type 1	B. malayi	XP_00189490	3e^{-78}	no	188	35.1	4	8.92
51	SR00682	BmFKBP59	B. malayi	XP_00190931	3e^{-57}	no	173	37.0	4	10.30
52	SR03139	Proteasome subunit alpha type 3	B. malayi	XP_00189410	2e^{-51}	no	166	73.5	9	18.00
53	SR01985	Proteasome subunit alpha type 1	B. malayi	XP_00189018	3e^{-74}	no	151	41.1	5	10.00
54	SS03222	20S proteasome alpha 5 subunit	B. malayi	XP_00189828	2e^{-74}	no	167	30.5	4	9.70
55	SR00809	Proteasome subunit beta type 2	B. malayi	XP_00189726	1e^{-71}	no	202	15.8	2	7.43
56	SR00774*	Ubiquitin	C. elegans	NP_741157	0.0	no	439	84.1	6	18.68
Nucleic acid metabolism										
57	SR01051*	Nucleoside di-phosphate kinase	B. malayi	XP_00190145	9e^{-70}	no	194	47.4	7	15.53
Structural proteins										
58	SS01554*	Actin 1	B. malayi	XP_00189795	3e^{-55}	no	376	38.6	11	24.80
59	SR01247*	Profilin	B. malayi	XP_00189578	8e^{-47}	no	132	81.8	7	17.28
60	SR00329	Yeast actin inter-acting protein 1	B. malayi	XP_00189936	2e^{-66}	no	210	38.1	7	15.45
61	SR00358	Actin depoly-merizing factor 1	B. malayi	XP_00189055	7e^{-64}	no	164	17.7	2	4.89
62	SS01430*	Tropomyosin	T. pseu-dospiralis	Q8WR63	4e^{-124}	no	284	34.9	10	24.18
Sugar-binding										
63	SR00857	Galectin protein 5	C. elegans	NP_495163	7e^{-60}	yes	183	28.4	4	9.54
64	SS00840	Galectin 1	T. circum-cinata	AAD39095	4e^{-127}	no	278	55.8	14	32.70
65	SR00900	Galectin	B. malayi	XP_00189648	2e^{-68}	no	163	54.6	6	17.50
66	SR00627	Galectin	H. contor-tus	AAF63405	7e^{-139}	no	276	38.0	10	25.56
Fatty acid binding										
67	SR00858*	Fatty acid binding protein	A. suum	P55776	7e^{-60}	yes	165	60.0	12	25.00
68	SR00605*	S1 protein	O. volvulus	CAA59101	2e^{-32}	no	180	31.7	6	13.81
69	SR00229*	Lipid binding pro-tein family member	C. elegans	NP_491928	7e^{-18}	no	116	38.8	5	14.10
Developmental processes										
70	SR00876	Calponin protein 3	C. elegans	O01542	3e^{-55}	no	144	76.4	13	31.21
71	SR01042	Elongation factor 2	C. elegans	P29691	0.0	no	531	29.2	12	33.34
72	SR00371	Nucleosome ass-embly protein 1	B. malayi	A8PJH0	2e^{-29}	no	247	42.1	8	16.00
73	SR02807	Elongation factor 1-beta/1-delta	B. malayi	XP_00189713	2e^{-55}	no	214	8.9	2	4.00
74	SS01391	Small subunit ribo-somal protein 28	P. maupasi	ABR87582	2e^{-25}	no	68	33.8	2	5.70

Table 1 a continued

Cluster		BLAST Alignment	Species	Accession Number	E	SP	EST Lgt.	% Cov.	# Pep.	UPS
Heat-shock proteins										
75	SR01060	Heat-shock protein 70	P. trichosuri	AAF87583	0.0	no	644	25.8	10	21.39
76	SR00952	Heat-shock protein 90	B. pahangi	CAA06694	$3e^{-137}$	no	338	26.0	7	16.59
77	SR00728	Heat-shock protein 60	S. ratti	ABY65231	0.0	no	563	64.1	33	71.01
78	SS01752	Chaperonin 10	S. ratti	ABN49241	$6e^{-51}$	no	109	33.9	3	6.17
79	SR00065	Heat-shock protein 70-C	H. glycines	AAM93256	$9e^{-89}$	yes	185	11.9	2	4.00
Cytoplasmatic										
80	SS01336*	Calexitin family member (CEX-1)	C. elegans	NP_495034	$4e^{-62}$	no	196	29.6	5	10.00
81	SR02058	Iron regulatory protein 1A	D. melanogaster	NP_477371	$9e^{-59}$	no	183	29.5	3	7.76
82	SS02442	Aconitase family member	C. elegans	NP_509898	$7e^{-72}$	no	166	17.5	2	6.79
83	SR02277	4-Hydroxyphenylpyruvate dioxygenase	C. elegans	NP_499324	$1e^{-86}$	no	199	58.8	6	16.14
84	SR00922*	14-3-3 family member	C. elegans	NP_509939	$4e^{-110}$	no	250	59.2	12	32.14
Other functions										
85	SR04614	Probable aspartate amino-transferase	C. elegans	Q22067	$5e^{-81}$	no	187	63.6	10	22.78
86	SR00060	Rab GDP dissociation inhibitor alpha, putative	B. malayi	EDP37909	$2e^{-66}$	trun	196	16.3	3	7.70
87	SS03041*	Conserved cysteine/glycine domain protein	C. elegans	NP_502842	$4e^{-33}$	yes	170	22.4	4	8.01
88	SS01143	Transthyretin-like protein	C. brenneri	ACD88894	$3e^{-48}$	trun	141	48.9	5	12.51
89	SR00146*	Transthyretin-related domain family member	C. elegans	NP_498657	$6e^{-33}$	yes	146	58.2	7	14.54
90	SR00027	Elongation factor 1-alpha	B. malayi	XP_001896880	0.0	no	455	29.5	9	22.79
91	SR00743	Glutathione S-transferase	A. suum	P46436	$3e^{-40}$	no	205	42.4	6	12.03
92	SR02533	Elongation factor 1	B. malayi	XP_001901841	$2e^{-33}$	no	183	42.1	5	10.00
93	SR01119	Nuclear transport factor 2	B. malayi	XP_001894897	$3e^{-47}$	no	132	48.5	5	10.01
94	SR01065	Glutamate dehydrogenase	B. malayi	XP_001893113	0.0	no	383	25.1	8	16.24
95	SR02773	LEThal family member	C. elegans	NP_498730	$2e^{-86}$	no	207	25.1	3	6.50
96	SR00950	Activated protein kinase C receptor	B. malayi	XP_001898740	$2e^{-131}$	no	326	30.1	6	13.80
97	SR02588	Inorganic pyrophosphatase	A. suum	BAC66617	$2e^{-57}$	yes	190	13.2	2	5.74

Table 1a continued

	Cluster	BLAST Alignment	Species	Accession Number	E	SP	EST Lgt.	% Cov.	# Pep.	UPS
98	SS00743	Transthyretin-related domain family member	C. elegans	NP_871961	$5e^{-42}$	yes	146	32.9	3	8.64
99	SR00387	Cyanate hydratase family protein	B. malayi	XP_001897006	$4e^{-32}$	no	171	37.4	5	11.08
100	SR00957*	SEC-2 protein	G. pallida	CAA70477	$4e^{-35}$	yes	181	28.2	3	6.25
101	SS00219	Glutamate dehydrogenase	H. contortus	AAC19750	$6e^{-68}$	no	154	16.9	2	4.59
102	SS00999	GABA transaminase family member	C. elegans	NP_501862	$1e^{-64}$	no	199	10.6	2	5.57
103	SR00756	ADP ribosylation factor 79F	D. melanogaster	NP_476955	$2e^{-100}$	no	181	43.6	5	11.79
104	SS00738	OV-16 antigen precursor	B. malayi	XP_001899662	$1e^{-61}$	no	173	21.4	2	7.88
105	SR02238	Submergence induced protein 2A	A. thaliana	AAM63708	$1e^{-20}$	no	120	33.3	3	7.44
106	SR02779	DVA-1 polyprotein precursor	D. viviparus	Q24702	$3e^{-07}$	yes	133	24.8	2	4.05
107	SS02521	Transthyretin-related domain family member	C. elegans	NP_499054	$6e^{-43}$	yes	143	21.7	2	4.77
108	SR00623	Lysozyme family member (lys-3)	C. elegans	NP_500206	$3e^{-47}$	yes	137	24.8	3	9.72
109	SR00930	Translationally controlled tumor protein like protein	S. ratti	ABF69532	$2e^{-74}$	no	170	28.2	5	15.81
110	SS01532	Gaba	B. malayi	XP_001900805	$6e^{-52}$	no	117	39.3	2	5.00
111	SR02279	Aldehyde dehydrogenase (alh-8)	C. elegans	NP_001022078	$2e^{-68}$	no	183	25	2	4.21
112	SR00351	Major sperm protein (msp-78)	C. elegans	NP_501742	$2e^{-64}$	no	127	21.3	2	4.03
113	SR00903	SXP-1 protein	L. loa	AAG09181	$4e^{-06}$	trun	320	13.8	3	6.93
114	SS03457	Eukaryotic initiation factor family member	C. elegans	NP_493272	$3e^{-31}$	no	85	57.6	3	6.49
Not assigned										
115	SR00866	Hypothetical protein CBG18957	C. briggsae	XP_001674360	$7e^{-139}$	yes	338	51.5	21	46.19
116	SR01871*	L3NieAg.01	S. stercoralis	AAD46493	$4e^{-21}$	trun	169	71.0	18	46.28
117	SR00795*	LL20 15 kDa ladder antigen	B. malayi	XP_001901667	$2e^{-32}$	no	250	41.2	8	16.14
118	SS00892*	Immunogenic protein 3	B. malayi	Q6S5M8	$2e^{-20}$	yes	127	22.0	3	8.18
119	SR00488*	Allergen polyprotein homolog	S. stercoralis	AAB97360	$1e^{-31}$	trun	205	22.4	4	11.09
120	SR00403	Similar to dj-1	A. mellifera	XP_624271	$1e^{-30}$	no	147	35.4	4	8.00
121	SR00753	T19B10.2	C. elegans	NP_505848	$1e^{-63}$	yes	191	37.7	5	10.00
122	SR02634	Hypothetical protein	A. aegypti	XP_001649379	1.3	no	172	24.4	3	7.32
123	SR00205	Hypothetical protein	C. elegans	XP_001667775	$8e^{-66}$	no	168	21.4	3	7.75
124	SR00281	F59A2.3	C. elegans	NP_497701	$5e^{-68}$	no	226	19.5	3	6.08

Table 1a continued

	Cluster	BLAST Alignment	Species	Accession Number	E	SP	EST Lgt.	% Cov.	# Pep.	UPS
125	SR00872	Hpothetical protein	C. briggsae	XP_001676876	$1e^{-134}$	no	307	38.8	9	22.47
126	SR00768	UPF0587	C. elegans	Q9BI88	$1e^{-43}$	no	169	18.9	3	7.22
127	SR00250	Hypothetical protein CBG03400	C. briggsae	XP_001670800	$5e^{-53}$	no	170	21.2	3	7.98
128	SR02918	Hypothetical protein CBG04059	C. elegans	XP_001668188	$6e^{-78}$	no	188	32.4	4	11.38
129	SR00573	Hypothetical protein CBG21098	C. briggsae	XP_001666425	$3e^{-34}$	yes	181	14.4	2	4.95
130	SR02535	Hypothetical 19.4 kDa protein ZC395.10	B. malayi	XP_001897485	$5e^{-05}$	no	119	29.4	2	4.04
131	SS01194	K02F2.2	C. elegans	NP_491955	$3e^{-134}$	no	260	26.5	5	10.14
132	SR00840*	Putative uncharacterized protein	C. briggsae	XP_001668392	$5e^{-49}$	yes	204	40.2	7	15.37
133	SR01280	Hypothetical protein CBG03385	C. briggsae	CAP24286	$2e^{-59}$	no	168	48.2	5	10.39
134	SR00428	Hypothetical protein F01G10.1	C. elegans	NP_501878	$1e^{-88}$	no	190	60.0	6	12.04

9.1.2 Table 1b: Nematode RefSeq proteins found in supernatants from the parasitic, the infective and the free-living stages

	Acc. Number	BLAST Alignment	Species	SP	Frag. Lgt.	% Cov.	# Pep.	UPS
135	gi|71996708	Glucose 6 phosphate isomerase (gpi-1)	C. elegans	no	586	26.6	11	25.00
136	gi|71995829*	Enolase family member (enol-1)	C. elegans	no	465	14.4	6	15.44
137	gi|17535107	Lactate dehydrogenase family member (ldh-1)	C. elegans	no	333	7.8	3	8.66
138	gi|71983429	C46F11.2b	C. elegans	no	473	9.1	3	6.00
139	gi|159183	Phosphoenolpyruvate carboxykinase	H. contortus	no	619	12.6	5	11.01
140	gi|17554310	Malate dehydrogenase family member (mdh-1)	C. elegans	no	341	20.2	4	9.16

9.1.3 Table 2a: List of *Strongyloides* EST cluster numbers found only in E/S products from infective larvae

	Cluster	BLAST Alignment	Species	Accession Number	E	SP	EST Lgt.	% Cov.	N^O Pep.	UPS
Oxidative metabolism										
141	SR03650	Superoxide dismutase	C. elegans	NP_492290	7e^{-60}	no	129	17.8	2	4.00
Carbohydrate metabolism										
142	SR00435	Pyruvate kinase	B. malayi	XP_001898629	4e^{-64}	no	176	16.4	2	4.00
143	SS00237	Carbohydrate phosphorylase	B. malayi	XP_001894669	1e^{-38}	no	111	54.9	4	8.00
144	SS02195	Carbohydrate phosphorylase	B. malayi	XP_001894669	1e^{-61}	no	148	39.1	4	6.32
145	SR00710	Trehalase family protein	B. malayi	XP_001900224	2e^{-51}	yes	281	11.0	2	4.01
146	SR01315	Phosphoenol-pyruvate carboxykinase	H. contortus	P29190	5e^{-34}	no	106	32	3	7.55
147	SS02356	Hexokinase family protein	B. malayi	XP_001897796	2e^{-16}	no	60	81.6	4	8.00
148	SR01676	Hexokinase family protein	B. malayi	XP_001897796	7e^{-68}	no	189	18.5	2	5.70
149	SR01818	UDP-Galactose 4´epimerase	B. malayi	XP_001902615	4e^{-20}	yes	81	60.5	2	4.00
Cytosol energy metabolism										
150	SS00138	Adenylate kinase	B. malayi	XP_001894222	4e^{-62}	no	149	49.6	6	14.00
151	SR00089	Aldehyde dehydrogenase	C. elegans	NP_498081	9e^{-76}	no	187	31.5	4	8.00
152	SR03043	Aldehyde dehydrogenase	C. elegans	NP_503467	2e^{-56}	no	162	33.3	4	8.00
153	SS03123	Aldehyde dehydrogenase	C. elegans	NP_00102212	4e^{-33}	no	134	25.3	2	4.54
154	SR03212	Aldehyde dehydrogenase	C. elegans	NP_498081	7e^{-76}	no	188	13.8	2	4.44
155	SR00700	Na, K-ATPase alpha subunit	C. elegans	AAB02615	3e^{-101}	no	233	29.2	4	9.71
156	SS01028	Na, K-ATPase alpha subunit	B. malayi	XP_001901816	4e^{-70}	no	166	14.5	2	4.00
157	SR00383	Propionyl Coenzyme A carboxylase	C. elegans	NP_509293	2e^{-17}	no	76	50.0	3	9.10
158	SR01758	2-oxoglutarate dehydrogenase	B. malayi	XP_001897753	6e^{-76}	no	185	20.5	2	5.60
159	SS00819	Citrate synthase family member	C. elegans	NP_499264	7e^{-61}	no	163	16.4	2	5.22
160	SR02784	Acyl-CoA binding protein	C. elegans	NP_509822	3e^{-36}	no	120	18.3	2	5.40
161	SS00766	Vacuolar H ATPase family member	C. elegans	NP508412	6e^{-98}	no	255	14.5	2	4.67
162	SS01243	Vacuolar H ATPase protein 16	B. malayi	XP_001901105	5e^{-177}	no	349	8.6	2	4.00
163	SS00854	GTP-binding nuclear protein RAN/TC4	B. malayi	XP_001900408	9e^{-95}	no	172	13.9	2	4.00

Table 2a continued

	Cluster	BLAST Alignment	Species	Accession Number	E	SP	EST Lgt.	% Cov.	N° Pep.	UPS
Protein digestion and folding										
164	SR11111	Metalloproteinase precursor	S. stercoralis	AAK55800	$2e^{-61}$	trun	265	43.8	6	13.72
165	SR04474	Peptidase family M1 containing protein	B. malayi	XP_001897028	$2e^{-64}$	no	196	32.1	4	10.36
166	SR02466	Calpain family member	C. elegans	NP_498741	$1e^{-76}$	no	180	20.6	2	4.00
167	SR02020	Calpain family protein 1	B. malayi	XP_001897507	$1e^{-53}$	no	171	36.3	3	6.00
168	SS01570	EF hand family protein	B. malayi	XP_001901225	$3e^{-83}$	no	176	26.1	2	4.00
169	SR02697	Zinc carboxypeptidase family protein	B. malayi	XP_001902361	$2e^{-44}$	yes	161	16.1	2	5.41
170	SR03248	Prolyl oligopeptidase	B. malayi	XP_001894227	$3e^{-38}$	no	190	8.9	2	5.15
Structural proteins										
171	SR01001	Myosin – filarial antigen	B. malayi	AAB35044	0.0	trun	490	41.2	13	13.67
172	SS02558	Disorganised muscle protein 1	B. malayi	XP_001899521	$2e^{-50}$	no	189	16.9	4	8.00
173	SR00587	Histone H2A	B. malayi	XP_001895511	$5e^{-59}$	no	129	24.8	2	5.10
174	SR04861	Protein unc-22	B. malayi	XP_001899339	$1e^{-32}$	no	193	30.0	3	6.82
175	SR03042	Uncoordinated family ember	C. elegans	NP_497044	$2e^{-48}$	yes	170	17.6	2	4.00
176	SR00848	Uncoordinated family member	C. elegans	NP_001021093	$2e^{-149}$	no	338	10.7	3	6.00
177	SS00698	Vinculin	B. malayi	XP_001899040	$8e^{-65}$	no	163	23.3	2	4.00
178	SS01450	Intermediate filament protein	A. suum	CAA60047	0.0	no	381	12.3	4	8.02
179	SR02037	Intermediate filament protein	B. malayi	XP_001900185	$1e^{-37}$	no	106	61.3	3	6.00
180	SR00426	Intermediate filament tail domain	B. malayi	XP_001901413	$1e^{-36}$	no	310	6.1	2	4.86
181	SS00348	Profilin	B. malayi	ACD47109	$9e^{-20}$	no	80	48.8	3	7.06
182	SS01266	Myosin-4	C. elegans	P02566	$3e^{-94}$	no	258	26.0	6	11.45
183	SR00998	Myosin light chain family member	C. elegans	NP_510828	$2e^{-73}$	no	170	31.8	4	10.60
184	SS01012	Myosin tail family member	B. malayi	XP_001901629	$1e^{-99}$	no	311	16.4	4	7.84
185	SS03393	Filamin/ABP280 repeat family protein	B. malayi	XP_001894616	$2e^{-45}$	no	157	19.7	2	5.72
186	SR05180	ERM-1	C. elegans	NP_491559	$2e^{-23}$	no	192	16.1	2	4.52
187	SR00624	Tubulin alpha chain	H. contortus	P50719	$1e^{-103}$	no	232	12.9	2	4.03
188	SR04696	Tubulin alpha chain	C. elegans	NP_496351	$3e^{-57}$	no	136	16.9	2	4.00
189	SR00547	Alpha-tubulin	B. malayi	XP_001893894	$1e^{-71}$	no	147	14.3	2	4.00
190	SS01136	Beta-tubulin	S. ratti	AAY16349	0.0	no	375	11.2	3	8.94
191	SR02940	Collagen alpha 2 subunit	A. suum	P27393	$3e^{-05}$	yes	150	18.7	2	4.05

Table 2a continued

	Cluster	BLAST Alignment	Species	Accession Number	E	SP	EST Lgt.	% Cov.	N° Pep.	UPS
192	SS02105	L-Plastin	B. malayi	NP_00189919	$1e^{-57}$	no	188	26.0	3	8.84
193	SR02137	L-Plastin	B. malayi	NP_00189919	$6e^{-66}$	no	177	16.3	2	5.41
194	SS02310	L-Plastin	B. malayi	NP_00189919	$7e^{-51}$	no	146	17.8	2	4.00
Fatty acid binding										
195	SR02741	Fatty acid retinoid binding protein	W. bancrofti	AAL33794	0.37	no	139	15.8	4	10.02
196	SS02274	Fatty acid retinol binding protein 1 precursor	W. bancrofti	Q9WT54	$7e^{-04}$	yes	155	27.7	4	8.56
Developmental processes										
197	SR03164	Reticulon protein family memeber	C. elegans	NP_506656	$5e^{-31}$	no	225	16.4	2	5.15
Heat-shock proteins										
198	SS01374	Small heat-shock protein 12.6	B. malayi	XP_00190059	$6e^{-13}$	no	153	16.9	2	4.00
Other functions										
199	SR01976	Prion like protein	B. malayi	XP_00190283	$3e^{-08}$	trun	78	43.6	3	6.00
200	SR00769	Prion like protein	B. malayi	XP_00190283	$4e^{-58}$	trun	213	15.5	2	4.78
201	SR01328	Class V aminotransferase	H. glycines	AAK26375	$3e^{-37}$	no	162	14.2	2	6.42
202	SS02304	ABC transporter family member	C. elegans	NP_503175	$2e^{-79}$	yes	182	26.3	4	8.60
203	SR02297	Tyrosinase like protein	C. elegans	Q19673	$7e^{-43}$	yes	192	20.8	3	6.00
204	SR01746	Acid phosphatase	B. malayi	XP_00190179	$2e^{-28}$	yes	190	19.4	2	4.00
205	SS02787	Dihydropyrimidinase family member	C. elegans	NP_501797	$2e^{-44}$	no	221	19.9	2	5.40
206	SR00906	RAS-related protein O-RAL	B. malayi	XP_00190054	$1e^{-76}$	yes	215	13.0	2	5.60
207	SS01511	RAS-related protein RAB-1A	B. malayi	XP_00190194	$7e^{-104}$	no	205	57.5	8	16.05
208	SR01803	Thiosulfate sulfurtransferase	B. malayi	XP_00190163	$1e^{-15}$	no	174	53.4	7	14.03
209	SS00527	Haloacid dehalogenase like hydrolase	B. malayi	XP_00189536	$1e^{-57}$	no	245	14.2	2	6.03
210	SS01895	Glutamate cysteine ligase modifier subunit	B. malayi	XP_00189995	$3e^{-29}$	no	182	20.9	3	7.57
211	SS01689	AvL3-1	A. viteae	AAA17420	$6e^{-63}$	no	195	12.3	2	4.00
212	SR03954	CAP protein	B. malayi	XP_00189188	$2e^{-53}$	no	242	28.1	4	9.22
213	SS00973	Ubiquitin E2 variant family member	C. elegans	NP_493578	$8e^{-53}$	no	141	29.0	3	8.02
214	SS00694	Trans thyretin related family member	C. elegans	NP_499054	$1e^{-40}$	yes	138	28.9	3	6.01
215	SS00453	Transthyretin-like family protein	B. malayi	XP_00189936	$3e^{-44}$	yes	125	17.6	2	4.00

Table 2a continued

	Cluster	BLAST Alignment	Species	Accession Number	E	SP	EST Lgt.	% Cov.	N° Pep.	UPS
216	SS01698	Patterned expression site family member	C. elegans	NP_506610	$3e^{-57}$	no	157	14.6	2	4.12
217	SS00110	Displaced gonad family member	C. elegans	NP_741200	$3e^{-30}$	yes	172	15.1	2	4.00
218	SR02558	Lethal family member (let-805)	C. elegans	NP_001022641	$1e^{-69}$	yes	190	37.8	5	13.52
219	SS02590	Sensory Axon guidance family member	C. elegans	NP_001033397	$3e^{-42}$	yes	169	59.7	6	12.00
220	SR04483	NCAM – neural cell adhesion molecule	C. elegans	NP_741708	$4e^{-05}$	yes	191	22.5	3	7.44
221	SR00292	Transmembrane matrix receptor MUP-4	C. elegans	AAK69172	$7e^{-50}$	yes	238	18.0	3	8.89
222	SS00025	Rab GDP dissociation inhibitor alpha	B. malayi	XP_001893230	$7e^{-73}$	no	169	15.5	2	4.49
223	SR00641	Vacuolar ATP synthase subunit B	B. malayi	XP_001893872	$6e^{-142}$	no	266	42.4	5	8.00
224	SS01502	Synaptobrevin	B. malayi	XP_001902083	$3e^{-32}$	no	111	18.0	2	4.00
225	SR03602	Nematode Kynurenine Aminotransferase	C. elegans	NP_510355	$9e^{-27}$	no	140	33.6	2	5.30
226	SS02977	Major allergen Ani s1	A. simplex	Q7Z1K3	$6e^{-22}$	yes	195	25.6	3	6.00
227	SR01398	Protein phosphatase 2C	B. malayi	XP_001898345	$3e^{-61}$	no	198	14.6	2	4.00
228	SS01495	RAS-related protein RAB-5B	B. malayi	XP_001898689	$9e^{-88}$	no	215	13.0	2	4.00
229	SS01382	Cyclase associated protein	B. malayi	XP_001891888	$3e^{-69}$	no	208	9.6	2	4.64
230	SS01764	DJ1 mammalian transcriptional regulator	C. elegans	NP_493696	$2e^{-32}$	no	161	12.6	2	4.38
231	SS00302	Methylamalonyl CoA epimerase family member	C. elegans	NP_492120	$3e^{-54}$	yes	153	21.3	3	8.17
232	SR03037	Trans thyretin related family member	C. elegans	NP_499054	$1e^{-30}$	yes	147	50.3	5	11.80
233	SS02764	Trans thyretin related family member	C. elegans	NP_509839	$2e^{-35}$	yes	128	24.2	3	6.06
234	SR00166	EF-1 guanine nucleotide exchange domain	B. malayi	XP_001897042	$1e^{-34}$	no	195	24.7	3	8.78
235	SR01252	RRN RNA binding domain family member	C. elegans	NP_497891	$2e^{-29}$	no	117	23.9	2	4.24
236	SS02951	Glutamate synthase	B. malayi	XP_001899642	$1e^{-47}$	no	169	21.8	3	6.77
237	SR02060	Cell division cycle related family member	C. elegans	NP_495705	$3e^{-92}$	no	193	32.6	4	9.50
238	SS03239	Cell division cycle related family member	C. elegans	NP_496273	$1e^{-92}$	no	206	17.9	2	4.06

Table 2a continued

	Cluster	BLAST Alignment	Species	Accession Number	E	SP	EST Lgt.	% Cov.	N° Pep.	UPS
239	SS01112	Translational endoplasmatic reticulum ATPase	B. malayi	XP_001902553	$3e^{-115}$	no	256	23.0	4	8.00
240	SR03221	Ubiquitin carboxyl terminal hydrolase	B. malayi	XP_001902802	$5e^{-55}$	no	178	21.9	3	6.00
241	SS01978	Aconitase 2	H. sapiens	AAH26196	$1e^{-68}$	no	172	20.8	2	4.60
242	SR00196	Threonyl t-RNA synthetase	B. malayi	XP_001893343	$1e^{-91}$	no	189	12.1	2	4.00
243	SS02820	Methylmalonyl CoA mutase homolog	C. elegans	NP_497786	$5e^{-94}$	no	195	11.7	2	4.00
244	SR01516	N-acetyl-galactosaminyl-transferase	C. elegans	AAC13673	$9e^{-88}$	yes	198	32.8	4	8.00
245	SR02185	Bifunctional aminoacyl-tRNA synthase	B. malayi	XP_001894567	$2e^{-76}$	no	187	14.9	2	5.17
246	SS00992	Receptor mediated endocytosis family member	C. elegans	NP_001024193	$1e^{-103}$	no	248	10.4	2	4.00
247	SR00606	NAD-dependent malic enzyme	B. malayi	XP_001902120	$2e^{-115}$	no	306	25.8	4	8.18
248	SR02855	NAD-dependent malic enzyme	B. malayi	XP_001902120	$5e^{-62}$	no	158	12.0	2	4.38
249	SS00946	Protein phosphatase PP2A regulatory subunit	B. malayi	XP_001894294	$8e^{-97}$	no	325	6.8	2	4.00
250	SR00901	TPR domain containing protein	B. malayi	XP_001902724	$5e^{-48}$	no	226	34.9	6	12.73
251	SS00837	B. malayi antigen	B. malayi	XP_001900036	0.09	yes	269	4.5	2	5.20
252	SR00660	Protein phosphatase methylesterase 1	B. malayi	XP_001896165	$3e^{-46}$	no	194	27.8	3	6.46
253	SS02248	Protein phosphatase methylesterase 1	B. malayi	XP_001896165	$2e^{-37}$	no	156	7.7	2	4.02
254	SS00421	Eukaryotic type carbonic anhydrase	B. malayi	XP_001901136	$8e^{-36}$	no	151	21.8	2	4.30
255	SS01166	Endophilin related protein	B. malayi	XP_001898188	$3e^{-89}$	no	277	8.3	2	5.71
256	SR00614	Leucine rich repeat family protein	B. malayi	XP_001896670	$5e^{-26}$	no	198	15.1	2	4.06
257	SR03119	Short chain reductase/dehydrogenase	B. malayi	XP_001900343	$1e^{-46}$	no	179	30.1	5	11.57
258	SR00255	Ubiquitin fusion degradation protein	B. malayi	XP_001902965	$5e^{-49}$	no	174	21.8	3	6.33
259	SR00141	C2 domain containing protein	B. malayi	XP_001895890	$3e^{-35}$	no	165	21.8	3	6.02
260	SR02207	Calponin 1	B. malayi	XP_001898596	$2e^{-57}$	no	145	22.0	2	5.70
261	SS01425	Ubiquitin ligase complex component	C. elegans	NP_492513	$1e^{-58}$	no	167	20.3	2	4.08
262	SS01049	RAS like GTP binding protein RhoA	B. malayi	XP_001896906	$2e^{-80}$	no	177	23.1	2	4.00
263	SR01305	Peptide methionine sulfoxide reductase msrB	B. malayi	XP_001900393	$2e^{-41}$	no	166	12.0	2	4.00

Table 2a continued

	Cluster	BLAST Alignment	Species	Accession Number	E	SP	EST Lgt.	% Cov.	N^O Pep.	UPS
264	SS00103	17.1 kDa polypeptide	B. malayi	XP_001896848	2e^{-19}	no	79	41.8	2	4.00
265	SS01330	Vacuolar ATP synthase subunit G	B. malayi	XP_001893897	1e^{-38}	no	125	21.6	2	4.00
266	SS02063	Vacuolar H ATPase family member	C. elegans	NP_496217	3e^{-49}	no	121	33.1	3	6.00
267	SR00694	High mobility group protein	B. malayi	NP_001900502	1e^{-74}	no	192	19.8	3	6.01
268	SS01022	High mobility group protein	B. malayi	NP_001900502	1e^{-81}	no	191	13.1	2	6.19
269	SS00814	Small subunit ribosomal protein 21	P. sp. 6 RS5101	ABR87532	5e^{-34}	no	90	48.9	2	5.71
270	SR03619	Seryl t-RNA synthetase family member	C. elegans	NP_501804	1e^{-51}	no	194	14.4	2	4.00
271	SR03176	Hexokinase	H. contortus	CAB40412	4e^{-72}	no	172	16.3	2	4.00
272	SR02972	Vacuolar H ATPase family member	C. elegans	NP_001023451	4e^{-72}	no	167	11.4	2	5.23
273	SR01338	Vacuolar ATPase subunit C family protein	B. malayi	XP_001899274	2e^{-16}	no	66	65.2	3	6.00

Not assigned

	Cluster	BLAST Alignment	Species	Accession Number	E	SP	EST Lgt.	% Cov.	N^O Pep.	UPS
274	SR00386	L3NieAg.01	S. stercoralis	AAD46493	0.15	trun	112	50.8	5	13.50
275	SR02886	Hypothetical 35.6 kDa protein	B. malayi	XP_001899587	1e^{-71}	no	180	55.0	7	14.10
276	SS01256	Hypothetical protein Bm1_13900	B. malayi	XP_001894244	1e^{-26}	yes	228	39.0	5	10.00
277	SS01225	GH10174	D. grimshawi	EDW04076	3e^{-04}	yes	223	15.2	2	4.43
278	SS02094	Hypothetical protein	B. malayi	XP_001899474	3e^{-31}	no	145	24.8	2	4.01
279	SR01443	Hypothetical protein	C. elegans	CAP26647	3e^{-05}	no	44	65.9	3	7.70
280	SR02296	Hypothetical protein	B. malayi	XP_001896771	1e^{-33}	yes	144	20.1	2	6.93
281	SR00103	Hypothetical protein CBG23797	C. elegans	CAP20556	6e^{-47}	yes	143	16.1	2	4.00
282	SR01197	Hypothetical protein C10G8.3	C. elegans	NP_504417	2e^{-11}	yes	166	30.7	4	8.47
283	SS01885	Y73B6BL.25	C. elegans	NP_500983	6e^{-23}	yes	196	18.4	3	6.02
284	SS00215	Hypothetical protein CBG04133	C. briggsae	XP_001667008	2e^{-11}	yes	204	14.2	2	4.00
285	SS02239	C32F10.8a	C. elegans	NP_001021021	4e^{-54}	no	140	16.4	2	4.65
286	SR01604	Hypothetical protein CBG08398	C. briggsae	XP_001671673	4e^{-18}	yes	171	26.9	2	4.00
287	SR03753	K02D10.1b	C. elegans	NP_498936	5e^{-30}	no	154	14.3	3	10.05
288	SS00711	F15B9.10	C. elegans	NP_001122920	5e^{-48}	yes	180	14.8	2	4.99
289	SS00357	Hypothetical protein CBG16364	C. briggsae	XP_001667068	8e^{-29}	no	132	15.2	2	4.00
290	SR01525	C30C11.4	C. elegans	NP_498868	2e^{-28}	no	211	11.4	2	4.73
291	SR01791	Y105E8A.19	C. elegans	NP_740947	3e^{-11}	no	188	16.5	2	5.52

Table 2a continued

	Cluster	BLAST Alignment	Species	Accession Number	E	SP	EST Lgt.	% Cov.	N° Pep.	UPS
292	SS00863	Hypothetical protein CBG21807	C. briggsae	XP_001672698	$1e^{-64}$	yes	156	16.0	2	4.00
293	SR05055	VF13D12L.3	C. elegans	NP_496500	$2e^{-59}$	no	189	47.1	3	7.40
294	SR01321	Hypothetical protein Bm1_36850	B. malayi	XP_001898817	$1e^{-12}$	no	176	43.2	5	10.00
295	SS01276	F55F3.3	C. elegans	NP_510300	$2e^{-104}$	no	317	12.0	3	9.17
296	SR00450	Y6B3B.5a	C. elegans	NP_001021801	$5e^{-15}$	yes	136	19.1	2	4.99
297	SR02530	Hypothetical protein CBG03344	C. briggsae	NP_001670753	$2e^{-31}$	no	125	24.8	2	4.01
298	SR03285	Hypothetical protein CBG04457	C. briggsae	XP_001671529	$1e^{-77}$	no	236	10.2	2	4.00
299	SR00182	Hypothetical protein LOC100127627	X. tropicalis	NP_001106450	$4e^{-42}$	no	180	16.7	2	4.04
300	SS00742	Temporarily assigned gene name tag-253	C. elegans	NP_741571	$9e^{-53}$	yes	188	17.6	2	4.00
301	SS01477	Hypothetical protein Bm1_23005	B. malayi	XP_001896058	$2e^{-14}$	yes	241	8.7	2	6.09
302	SR00244	Y53H1B.1	C. elegans	NP_492900	$1e^{-58}$	yes	284	8.5	2	4.00
303	SS01439	ZK1307.1a	C. elegans	NP_001022509	$2e^{-79}$	no	237	18.6		6.91
304	SR01140	Hypothetical protein CBG23403	C. briggsae	XP_001673594	$6e^{-15}$	no	104	31.7	2	4.21
305	SR01943	Novel protein similar to COG3	D. rerio	CAQ14213	0.76	no	168	29.2	3	7.75
306	SS01523	Y54G2A.2a	C. elegans	NP_001023492	0.0	no	544	6.4	3	6.22
307	SR02095	Hypothetical protein CBG10437	C. briggsae	NP_001672867	$5e^{-36}$	no	119	24.4	2	4.10
308	SR00123	Unnamed protein product	T. nigroviridis	CAG11397	$3e^{-27}$	no	138	25.3	2	4.00
309	SR00366	hypothetical protein DDBDRAFT_0217849	D. discoideum AX4	XP_642992	$5e^{-08}$	no	190	40.5	6	13.15
310	SR02793	RO5H10.7	C. elegans	NP_001022271	$1e^{-15}$	no	212	12.7	2	5.17

9.1.4 Table 2b: Nematode RefSeq proteins only found in E/S products from the infective larvae

	Acc. Number	BLAST Alignment	Species	SP	Frag. Lgt.	% Cov.	# Pep.	UPS
311	gi|71987720	Lethal family member (let-805)	C. elegans	yes	4450	0.7	2	5.46
312	gi|17569483	Spectrin family member (sps-1)	C. elegans	no	2427	7.4	15	34.98
313	gi|17535441	R05F9.6	C. elegans	no	568	6.0	3	6.08
314	gi|17543050	Y37A1B.5	C. elegans	no	471	6.8	2	5.40
315	gi|17534675	G-protein, beta subunit (gpb-1)	C. elegans	no	340	7.4	2	4.05
316	gi|25151898	Dense body family member (deb-1)	C. elegans	no	999	2.5	2	6.09
317	gi|159893	Major body wall myosin	O. volvulus	no	1957	2.6	3	6.42
318	gi|72003683	Lin-5 interacting protein family member (lfi-1)	C. elegans	no	2350	0.8	2	4.58
319	gi|17534447	Aldehyde dehydrogenase family member (alh-6)	C. elegans	no	563	6.9	3	6.17
320	gi|25146553	C37H5.6a	C. elegans	no	457	4.2	2	4.08
321	gi|17507969	Homogentisate oxidase family member (hgo-1)	C. elegans	no	437	6.9	2	4.03
322	gi|71981209	C32F10.8a	C. elegans	no	504	5.8	2	4.13
323	gi|17568413	GPD family member	C. elegans	no	341	10.9	3	5.74
324	gi|17560440	F32D1.5	C. elegans	no	358	7.5	2	4.10
325	gi|74763811	Tubulin beta chain	T. trichiura	no	444	52.9	24	51.84
326	gi|32566457	Filamin (flna-1)	C. elegans	no	2747	0.8	2	5.95
327	gi|392788	Intermediate filament protein	B. malayi	no	506	6.1	4	8.13
328	gi|17570289	W07E11.1	C. elegans	no	2207	1.4	2	7.38
329	gi|17555418	Uncoordinated family member (unc-116)	C. elegans	no	815	2.9	2	4.97
330	gi|17541896	Isoleucyl t-RNA synthetase family member (irs-1)	C. elegans	no	1141	1.8	2	4.02
331	gi|38016557	UNC-18	H. contortus	no	588	3.9	2	7.65
332	gi|17554896	T04A8.7a	C. elegans	no	681	3.5	2	4.00
333	gi|17556182	Y54F10AR.1	C. elegans	no	336	10.7	2	4.00
334	gi|17508491	Proteasome alpha subunit family member (pas-4)	C. elegans	no	153	25.5	2	5.17
335	gi|17508451	Neuronal calcium sensor family member (ncs-2)	C. elegans	no	190	57.3	7	14.02
336	gi|25153953	Ubiquitin conjugating enzyme family member (ubc-13)	C. elegans	no	151	16.6	2	4.57

9.1.5 Table 3a: List of *Strongyloides* EST cluster numbers found only in E/S products from parasitic females

Cluster	BLAST Alignment	Species	Accession Number	E	SP	EST Lgt.	% Cov.	N° Pep.	UPS
Anti-oxidants									
337 SR00399	Thioredoxin family protein	B. malayi	XP_001892562	1e^{-43}	no	152	15.8	2	5.15
Carbohydrate Metabolism									
338 SR02118	Phosphoribosyl transferase	B. malayi	XP_001895434	2e^{-27}	no	152	31.6	3	6.24
339 SR02411	Glycosyl hydrolase	B. malayi	XP_001901753	3e^{-65}	yes	181	16.0	2	4.17
340 SS02201	Glycosyl transferase	B. malayi	XP_001901721	3e^{-99}	no	218	11.0	2	5.22
341 SS00512	Glucosidase beta subunit	B. malayi	XP_001899874	4e^{-37}	no	137	12.4	2	4.00
Protein digestion and folding									
342 SR01641	Prolyl endopeptidase	T. denticola	NP_971802	2e^{-12}	no	126	31.0	4	11.26
343 SR03191	Prolyl endopeptidase	R. norvegicus	EDL99674	3e^{-19}	no	189	51.3	6	16.69
344 SR03587	Metalloproteinase	N. vitripennis	XP_001606489	9e^{-06}	yes	166	42.8	6	14.23
345 SR03310	mp1	O. volvulus	AAV71152	2e^{-13}	yes	189	39.2	5	11.64
346 SR03901	Aspartyl protease (asp-2)	C. elegans	NP_505384	4e^{-43}	no	191	56.5	5	13.20
347 SR00074	Aspartyl protease (asp-2)	C. elegans	NP_505384	1e^{-59}	no	222	11.7	2	4.70
348 SR00564	Calumenin	C. elegans	NP_001024806	2e^{-134}	yes	286	19.2	4	10.87
349 SS00575	Calcyclin binding protein	T. castaneum	XP_967766	8e^{-19}	no	173	12.7	2	5.27
350 SR02954	Peptidase family M1 containing protein	B. malayi	XP_001901798	5e^{-27}	yes	171	14.0	2	4.00
351 SR02663	Metalloproteinase precursor	S. stercoralis	AAK55800	6e^{-12}	trun	182	19.2	3	6.91
352 SS00365	Ubiquitin family protein	B. malayi	XP_001902395	6e^{-14}	no	168	16.7	2	4.80
Structural proteins									
353 SR00945	Troponin T family member	C. elegans	NP_001024703	3e^{-103}	no	357	6.2	2	4.01
354 SR01499	Troponin family protein	B. malayi	XP_001898461	5e^{-50}	no	257	14.0	3	6.19
355 SS03220	Intermediate filament protein (ifa-3)	C. elegans	NP_510649	6e^{-38}	no	141	24.1	3	6.23
Heat-shock proteins									
356 SR00984	Small heat-shock protein	T. spiralis	ABJ55914	2e^{-21}	no	160	35.6	4	10.67
357 SR03349	Heat-shock protein HSP17	C. elegans	NP_001023958	2e^{-20}	no	157	39.5	5	10.04
358 SS01231	Heat-shock 70 kDa protein	B. malayi	XP_001892998	2e^{-20}	no	90	16.7	2	4.00

Table 3a continued

Cluster		BLAST Alignment	Species	Accession Number	E	SP	EST Lgt.	% Cov.	N° Pep.	UPS
Cytoplasmatic										
359	SR00396	Endoplasmin	B. malayi	XP_001899398	$7e^{-98}$	yes	231	16.5	3	8.56
360	SS00604	Endoplasmin	B. malayi	XP_001899398	$7e^{-54}$	yes	179	11.2	2	4.49
Nucleic acid metabolism										
361	SR01055	Ribosomal protein (rpl-4)	C. elegans	NP_491416	$1e^{-132}$	no	341	9.1	2	4.41
362	SR01073	Ribosomal protein (rpl-5)	C. elegans	NP_495811	$2e^{-119}$	no	290	19.3	4	9.47
363	SR00991	Small subunit ribosomal protein 1	P. sp. 3 CZ3975	ABR87454	$5e^{-115}$	trun	257	21.0	3	6.03
364	SR01039	Ribosomal protein L8 CG1263-PA	A. mellifera	XP_393671	$2e^{-112}$	no	243	8.6	2	4.04
365	SR01021	Putative ribosomal protein L15	D. citri	ABG81972	$4e^{-85}$	no	204	10.3	2	4.00
366	SR01047	Ribosomal protein (rps-3)	C. elegans	NP_498349	$3e^{-100}$	no	242	8.7	2	5.57
367	SS01567	40S ribosomal protein S8	B. malayi	XP_001893693	$2e^{-91}$	no	209	20.1	3	6.00
368	SR01072	Ribosomal protein (rpl-13)	C. elegans	NP_00102017	$9e^{-69}$	no	213	13.1	2	5.62
369	SR01036	Ribosomal protein L6	S. papillosus	ABK55146	$4e^{-56}$	trun	204	15.2	2	6.16
370	SS01551	40S ribosomal protein S9	B. malayi	XP_001894478	$1e^{-85}$	no	192	8.9	2	4.14
371	SR00979	Ribosomal protein L9	S. papillosus	ABK55147	$2e^{-73}$	trun	166	21.7	2	6.47
372	SS01553	60S ribosomal protein L11	B. malayi	XP_001892371	$1e^{-88}$	no	197	16.2	2	5.42
373	SR01004	Ribosomal protein L18e	T. castaneum	XP_968042	$6e^{-60}$	no	187	15.1	2	4.68
374	SR01002	Ribosomal protein (rps-18)	C. elegans	NP_502794	$5e^{-75}$	no	154	15.6	2	6.64
375	SR00943	Ribosomal protein (rpl-14)	C. elegans	NP_492576	$2e^{-36}$	no	134	20.9	2	4.07
Other function										
376	SS00866	EF hand family protein	B. malayi	XP_001901161	$2e^{-29}$	yes	177	13.6	3	6.01
377	SR01608	EF hand family protein	B. malayi	XP_001901161	$2e^{-37}$	yes	158	69.0	8	23.28
378	SR04713	Surface antigen BspA-like	T. vaginalis	XP_001315000	5.3	no	55	89.1	3	7.27
379	SR01156	Small nuclear ribonucleo-protein E	B. malayi	XP_001894662	$5e^{-24}$	no	88	33.0	2	5.40
380	SR04847	Acetylcholinesterase 2	D. destructor	ABQ58116	$1e^{-44}$	no	192	36.5	5	12.4
381	SR02054	Scavenger receptor cysteine rich protein	C. pipiens quinquefasciatus	XP_001866937	$3e^{-24}$	yes	328	14.6	4	10.44
382	SR04723	Abnormal nuclease family member	C. elegans	NP_509604	$8e^{-39}$	yes	187	20.3	2	4.84

Table 3a continued

	Cluster	BLAST Alignment	Species	Accession Number	E	SP	EST Lgt.	% Cov.	N° Pep.	UPS
383	SR01007	40S ribosomal protein S4	B. malayi	XP_00189272	$1e^{-117}$	no	262	7.6	2	5.51
384	SR01297	Immnunosuppressive ovarian message protein	A. suum	CAK18209	$2e^{-17}$	yes	324	19.8	4	8.15
385	SR03217	Immnunosuppressive ovarian message protein	A. suum	CAK18210	$6e^{-16}$	yes	193	10.9	2	4.00
386	SR00986	60S ribosomal protein L10	B. malayi	XP_00189297	$2e^{-85}$	no	189	17.5	3	8.26
387	SR00664	Pheromone biosynthesis-activating neuropeptide	H. virescens	AAO20095	0.058	yes	229	15.3	2	4.19
388	SS03344	SPARC precursor	B. malayi	XP_00189784	$2e^{-74}$	yes	175	20.6	3	6.23
389	SR04821	PAZ domain containing protein	B. malayi	XP_00189598	$8e^{-08}$	no	175	12.6	2	5.40
390	SR01592	Ankyrin repeat domain containing protein	M. musculus	CAQ51694	3.5	no	166	22.9	3	6.00
391	SR00441	SELD-1	C. elegans	NP_502604	$4e^{-54}$	no	179	15.1	2	5.40
392	SS01458	Beta-NAC-like protein	B. malayi	XP_00190247	$1e^{-59}$	no	156	20.5	2	4.01
393	SR04058	Nuclear movement protiein	B. malayi	XP_00189248	$4e^{-53}$	no	221	9.4	2	4.01
394	SR03161	Adenine phosphoribosyltransferase	L. vestofoldensis	ZP_01000391	2.0	no	65	36.9	2	4.00
395	SR01308	Isochorismatase family protein	B. malayi	XP_00190176	$9e^{-19}$	no	117	29.1	2	5.70
Not assigned										
396	SR04455	Hypothetical protein CBG05204	C. briggsae	XP_00166481	2.0	yes	88	56.8	5	10.64
397	SR03561	Hypothetical protein CBG05204	C. briggsae	XP_00167719	$1e^{-49}$	yes	150	32.7	3	6.00
398	SS00845	C05D11.1	C. elegans	NP_00102145	$1e^{-46}$	no	209	11.0	2	4.15
399	SR03259	Hypothetical protein	B. malyi	XP_00189639	$1e^{-64}$	no	171	25.7	3	6.00
400	SR03153	Hypothetical protein CBG19978	C. briggsae	XP_00167097	$9e^{-82}$	no	192	15.6	2	5.70
401	SS03048	Hypothetical protein CBG09854	C. briggsae	XP_00166547	$3e^{-53}$	yes	175	13.1	2	4.04
402	SR02153	Hypothetical protein CBG20335	C. briggsae	CAP37373	$2e^{-34}$	no	178	13.5	2	6.34
403	SR01296	Hypothetical protein CBG07632	C. briggsae	XP_00167777	$5e^{-59}$	yes	179	8.4	2	4.00
404	SR03570	Predicted protein	N. vectensis	XP_00163220	$1e^{-3}$	no	33	72.7	2	5.22
405	SR00696	T20D3.2	C. elegans	NP_501640	$7e^{-19}$	yes	225	14.7	2	4.00
406	SR02281	Hypothetical protein CBG23452	C. briggsae	XP_00167351	$2e^{-26}$	no	112	28.6	2	5.40

9.1.6 Table 3b: Nematode RefSeq proteins only found in E/S products from the parasitic females

	Acc. Number	BLAST Alignment	Species	SP	Frag. Lgt.	% Cov.	№ Pep.	UPS
407	gi|22759003	Acetylcholinesterase 1	N. americanus	yes	594	1.2	2	4.02
408	gi|6754388	Intelectin 1	M. musculus	yes	313	15.7	4	11.74
409	gi|3182894	Actin	B. malayi	no	376	22.3	5	15.55
410	gi|71992775	Y105E8A.19	C. elegans	no	722	3.0	2	4.55
411	gi|17534771	Heat-shock protein (hsp-4)	C. elegans	yes	657	4.6	3	7.57
412	gi|71986328	F48E8.3	C. elegans	yes	493	4.5	2	4.02
413	gi|20163188	Aldolase	H. glycines	no	366	7.4	2	4.01
414	gi|17508209	Acylsphingosine amidohydrolase	C. elegans	yes	393	9.4	2	5.06
415	gi|40388674	14-3-3b protein	M. incognita	no	251	13.1	3	7.70

9.1.7 Table 4a: List of *Strongyloides* EST cluster numbers found in E/S products from infective larvae and parasitic females

Cluster		BLAST Alignment	Species	Accession Number	E	SP	EST Lgt.	% Cov.	№ Pep.	UPS
Anti-oxidants										
416	SS00590	Glutathione peroxidase	*H. contortus*	AAT28332	$3e^{-74}$	no	197	25.9	4	8.00
417	SS01859	Peroxiredoxin	*C. elegans*	NP_497892	$4e^{-55}$	no	132	22.7	2	7.22
418	SR01250	Glutaredoxin	*D. rerio*	NP_001005950	$1e^{-34}$	no	177	13.6	2	4.02
Carbohydrate Metabolism										
419	SR05118	Fructose-1,6-biphosphatase	*C. elegans*	CAB69047	$9e^{-21}$	no	65	30.8	2	4.00
420	SR04547	Glycogenin-1	*B. malayi*	XP_001894812	$3e^{-41}$	no	135	50.4	4	9.73
421	SR03154	Fructose-1,6-biphosphatase	*C. elegans*	CAB69047	$6e^{-51}$	no	158	26.6	3	7.22
422	SS00993	UDP-glucose pyrophosphorylase	*B. malayi*	XP_001899721	$6e^{-71}$	no	209	20.6	3	7.58
423	SR03282	Phosphoglucomutase	*B. malayi*	XP_001894605	$8e^{-60}$	no	186	24.7	3	7.54
424	SS02308	Deoxiribose-phosphate aldolase	*C. elegans*	Q19264	$7e^{-42}$	no	200	20.0	3	6.00
Cytosol energy metabolism										
425	SR00526	Glyceraldehyde 3-phosphate dehydrogenase	*C. elegans*	NP_508534	$4e^{-154}$	no	318	33.3	7	15.19
426	SR00087	2-oxoglutarate dehydrogenase E1 component	*C. briggase*	Q623T0	$1e^{-88}$	no	190	26.3	5	10.00
427	SS03262	Dynein light chain 1	*B. malayi*	XP_001897484	$7e^{-44}$	no	89	27.0	2	6.22
428	SR00394	Putative sarcosine oxidase	*C. elegans*	Q18006	$2e^{-26}$	no	151	35.8	3	6.55
Protein digestion and folding										
429	SS01380	Serine carboxypeptidase	*C. elegans*	P52717	$1e^{-101}$	yes	247	11.7	2	4.05
430	SS02084	Leucyl Aminopeptidase	*C. briggsae*	CAP35649	$4e^{-28}$	no	190	16.8	3	7.49
431	SS00263	Proteasome subunit alpha	*B. malayi*	XP_001892613	$6e^{-101}$	no	250	30.8	6	14.07
432	SS00433	Protein disulfide isomerase A6	*B. malayi*	XP_001898144	$3e^{-77}$	yes	215	15.8	2	4.00
433	SR01918	Ubiquitin carboxyl terminal hydrolase	*B. malayi*	XP_001893306	$7e^{-34}$	no	180	23.3	3	7.76
434	SR02675	Aminopeptidase	*B. malayi*	XP_001901574	$2e^{-30}$	no	171	61.4	6	12.95
435	SS01197	Proteasome alpha subunit 2	*C. briggsae*	CAP28928	$5e^{-115}$	no	231	35.5	7	15.68
436	SS01557	Proprotein convertase 2	*B. malayi*	XP_001901951	$1e^{-112}$	yes	257	13.2	3	6.01
437	SS03260	Proteasome subunit alpha	*B. malayi*	XP_001901359	$3e^{-59}$	no	150	60.7	5	10.01
438	SS03110	Proteasome subunit beta	*B. malayi*	XP_001896421	$8e^{-43}$	no	120	25.8	2	4.00

Table 4a continued

	Cluster	BLAST Alignment	Species	Accession Number	E	SP	EST Lgt.	% Cov.	№ Pep.	UPS
439	SR00910	Proteasome alpha subunit-3	C. briggsae	CAP30116	$5e^{-109}$	no	238	31.9	5	10.07
440	SS03014	Proteasome alpha subunit-7	C. briggsae	XP_001678494	$5e^{-42}$	no	128	14.1	2	5.77
441	SS02746	Proteasome beta subunit-5	C. briggsae	CAP36972	$9e^{-35}$	no	94	36.2	2	4.00
442	SR00725	Proteasome beta subunit	C. elegans	NP_500125	$2e^{-57}$	no	171	26.3	3	7.53
Structural proteins										
443	SS03469*	Uncoordinated family member (unc-60)	C. elegans	NP_503427	$7e^{-56}$	no	144	47.9	5	11.05
444	SR00742	Uncoordinated family member (unc-52)	C. briggsae	CAP30283	$9e^{-57}$	trun	301	22.6	5	11.97
445	SS02042	Uncoordinated family member (unc-52)	C. briggsae	CAP30283	$8e^{-52}$	trun	171	16.4	2	4.02
446	SS03183	Two 7-bladed beta propeller domain	B. malayi	XP_001899636	$4e^{-93}$	no	218	30.7	4	10.32
447	SS01235	Histone H4	B. malayi	XP_001898784	$5e^{-50}$	no	103	31.1	3	6.80
448	SS00585	Histone (his-41)	C. elegans	NP_505464	$6e^{-47}$	no	123	19.5	2	4.57
449	SR01515	Myosin light chain 3	C. briggsae	CAP20750	$1e^{-53}$	no	151	60.9	11	24.63
Fatty acid binding										
450	SR01165	Lipid binding protein LBP-3	C. briggsae	CAP32783	$2e^{-06}$	yes	113	32.7	3	6.89
Developmental processes										
451	SR02808	Thioredoxin reductase	B. malayi	XP_001898729	$2e^{-32}$	yes	203	23.6	4	11.80
Cytoplasmatic										
452	SS00759	Fumarylacetoacetase	G. gallus	XP_413855	$2e^{-67}$	no	193	25.9	2	7.82
453	SS00416	Alanine aminotransferase	A. pisum	XP_001948711	$3e^{-30}$	no	119	25.2	2	4.04
454	SS00697	Aspartate aminotransferase	C. elegans	Q22067	$5e^{-56}$	no	165	16.4	2	4.42
455	SS03132	Ubiquitin like protein SMT3	B. malayi	XP_001900504	$1e^{-28}$	no	107	29.0	2	4.04
Other functions										
456	SS01841	Dihydrolipoyl dehydrogenase	B. malayi	XP_001896712	$7e^{-85}$	no	172	39.5	4	11.63
457	SS00317	Rab GDP dissociation inhibitor alpha	B. malayi	XP_001893230	$7e^{-77}$	no	183	33.3	4	9.70
458	SR00620	Elongation factor 1 gamma	B. malayi	XP_001901841	$2e^{-76}$	no	185	28.1	4	8.13
459	SS02569	P25-alpha family protein	B. malayi	XP_001901015	$5e^{-63}$	no	178	23.6	3	8.56
460	SR02760	Aldehyde dehydrogenase (alh-9)	C. elegans	NP_498263	$2e^{-80}$	no	189	22.8	3	7.30
461	SS00729	Lipoamide dehydrogenase	A. suum	AAD30450	$4e^{-79}$	no	190	28.9	5	10.69

Table 4a continued

Cluster		BLAST Alignment	Species	Accession Number	E	SP	EST Lgt.	% Cov.	№ Pep.	UPS
462	SS02079	Rab GDP dissociation inhibitor alpha	B. malayi	XP_001893230	$1e^{-15}$	no	50	40.0	2	4.00
463	SS01500	Ras-related protein Rab-11b	B. malayi	XP_001902705	$2e^{-104}$	no	213	36.6	7	19.12
464	SS01469	Aspartyl protease inhibitor	T. colubriformis	P59704	$3e^{-39}$	yes	257	10.9	3	6.18
465	SR03132	Esterase D	B. malayi	XP_001894981	$3e^{-57}$	no	180	43.9	7	15.71
466	SR00382	FAD-dependent oxidoreductase	B. malayi	XP_001892448	$6e^{-47}$	yes	181	11.0	2	5.84
467	SS01263	Aminotransferase	H. glycines	AAK26375	$1e^{-83}$	yes	232	26.7	4	8.28
468	SR02462	Enolase phosphatase E1	D. rerio	Q6GMI7	$2e^{-30}$	no	130	37.7	4	8.33
469	SS00496	TFG-1 protein	C. briggsae	CAP36192	$3e^{-10}$	no	168	26.8	3	6.38
470	SR00939	L3ni51	D. viviparus	AAT02162	$5e^{-23}$	yes	174	31.6	7	13.40
471	SR02164	Eukaryotic initiation factor	C. elegans	NP_493272	$1e^{-68}$	no	169	16.6	2	5.15
472	SR05203	Gut esterase 1	C. briggsae	Q04456	$7e^{-44}$	yes	188	15.4	2	5.06
473	SS02291	Hypothetical UPF0185 protein	B. malayi	XP_001900717	$8e^{-37}$	no	87	57.5	3	6.06
474	SR00981	Ribosomal protein, small subunit family member	C. elegans	NP_497978	$1e^{-91}$	no	259	16.6	2	5.19
475	SS02895	Hypothetical UPF0160 protein	B. malayi	XP_001901763	$4e^{-39}$	no	168	18.5	2	5.00
476	SS02425	Trans-thyretin related family member	C. elegans	NP_505304	$6e^{-23}$	yes	135	24.4	3	6.41
477	SR04322	Small nuclear ribonucleoprotein G	B. malayi	XP_001893831	$1e^{-19}$	no	69	26.1	2	4.20
478	SS01459	Macrophage migration inhibitory factor	S. ratti	ACH88456	$6e^{-64}$	no	123	24.4	2	5.73
479	SS03348	UBX-domain containing protein	C. briggsae	CAP32645	$4e^{-11}$	trun	103	35.9	3	6.00
480	SR00132	Aldehyde dehydrogenase (alh-9)	C. elegans	NP_498263	$6e^{-63}$	no	145	21.4	3	6.48
481	SR01418	Insulin degrading enzyme	B. malayi	XP_001896776	$9e^{-33}$	no	192	15.1	2	6.65
482	SS01558	Temporarily assigned gene name family member	C. elegans	NP_001022799	$2e^{-159}$	no	300	17.7	4	10.08
483	SS01411	EGF-like domain containing protein	C. elegans	O44443	$2e^{-48}$	yes	261	8.0	2	4.46
484	SS01462	Intracellular globin	S. trachea	AAL56427	$9e^{-37}$	no	152	21.7	2	4.14
485	SS01542	Ferritin	L. vannamei	AAX55641	$4e^{-40}$	no	182	17.6	2	5.40
486	SR05189	Early B-cell factor 3 (EBF3-S)	B. malayi	XP_001901522	$3e^{-20}$	no	188	11.2	2	4.00
487	SS00202	Pterin carbinol-amin dehydratase	C. elegans	NP_491982	$1e^{-31}$	no	107	50.5	4	8.00
488	SR01834	Proline synthethase associated protein	E. dispar	XP_001735540	$2e^{-23}$	no	185	10.3	2	5.40

Table 4a continued

Cluster		BLAST Alignment	Species	Accession Number	E	SP	EST Lgt.	% Cov.	№ Pep.	UPS
Not assigned										
489	SS00514	Hypothetical protein Bm1_00020	B. malayi	XP_001891532	$6e^{-28}$	yes	165	13.9	2	6.27
490	SS01162	Hypothetical protein CBG16927	C. briggsae	XP_001677229	$5e^{-56}$	yes	159	27.0	4	9.27
491	SR00325	Hypothetical protein CBG07481	C. briggsae	XP_001677874	$4e^{-41}$	no	159	17.0	2	4.00
492	SR04443	Hypothetical protein CBG00142	C. briggsae	XP_001677295	$3e^{-16}$	no	74	73.0	4	8.04
493	SS01064	F20D1.3	C. elegans	NP_510486	$8e^{-36}$	no	182	34.1	4	8.00
494	SR02739	Y39G8B.1b	C. elegans	NP_496924	$2e^{-63}$	no	223	34.1	5	10.36
495	SR05083	Y105C5B.15	C. elegans	NP_502904	$7e^{-17}$	yes	178	36.0	4	8.00
496	SR02187	Hypothetical protein CBG07632	C. briggsae	XP_001677777	$2e^{-59}$	yes	182	26.4	4	11.15
497	SS01940	Hypothetical protein CBG03955	C. briggsae	XP_001668096	$4e^{-43}$	no	110	33.6	2	5.35
498	SR00699	Y43F4B.5a	C. elegans	NP_001022872	$1e^{-91}$	no	239	20.1	3	9.53
499	SR01216	GH11603	D. grimshawi	XP_001989219	$2e^{-29}$	no	77	36.4	2	6.27
500	SR00248	Hypothetical protein CBG01313	C. briggsae	XP_001673156	$2e^{-109}$	no	234	26.1	4	9.97
501	SR00591	Hypothetical protein	B. malayi	XP_001900680	$4e^{-38}$	no	167	16.8	2	4.90
502	SS03444	Hypothetical protein F13H8.7	C. elegans	AAK31588	$7e^{-76}$	no	201	11.1	2	4.40
503	SR02678	Hypothetical protein F08F8.4	C. elegans	T15978	$1e^{-71}$	no	195	25.6	2	4.04

9.1.8 Table 4b: Nematode RefSeq proteins only found in E/S products from infective larvae and parasitic females

	Acc. Number	BLAST Alignment	Species	SP	Frag. Lgt.	% Cov.	№ Pep	UPS
504	gi|9972772	Catalase	A. suum	no	541	9.6	3	7.70
505	gi|17564388	T19B10.2	C. elegans	yes	368	8.4	2	6.00
506	gi|71997207	Variable abnormal morphology member	C. elegans	no	4.955	0.8	3	7.68
507	gi|37681504	Independent phosphoglycerate mutase	B. malayi	no	515	4.7	2	4.08
508	gi|71981858	C44E4.3	C. elegans	no	419	4.3	2	4.11
509	gi|63029700	Alpha tubulin	O. volvulus	no	450	11.8	3	6.03

9.1.9 Table 5a: List of *Strongyloides* EST cluster numbers found only in E/S products from the free-living stages

	Cluster	BLAST Alignment	Species	Accession Number	E	SP	EST Lgt.	% Cov.	№ Pep.	UPS
Protein digestion and folding										
510	SR00267	Carboxypeptidase	*S. coelicolor*	NP_630267	0.89	yes	243	11.5	2	4.00
511	SR02550	Putative serine protease F56F10.1	*C. elegans*	P90893	$2e^{-35}$	yes	239	11.7	2	5.22
512	SR05257	Putative serine protease F56F10.1	*C. elegans*	P90893	$2e^{-24}$	yes	185	25.4	2	5.52
513	SR02579	Metalloproteinase	*N. americanus*	ACB13196	$3e^{-46}$	yes	192	18.8	2	4.00
514	SR01063	Aspartyl protease precursor	*C. briggsae*	CAP30637	$2e^{-98}$	yes	359	10.9	2	4.49
Carbohydrate metabolism										
515	SR00479	Hexosaminidase B	*P. troglodytes*	XP_517705	$2e^{-46}$	yes	167	16.8	2	7.13
Heat-shock proteins										
516	SS01082	Hypothetical 86.9 kDa protein	*B. malayi*	XP_001896095	$5e^{-51}$	no	309	6.8	2	5.22
Developmental processes										
517	SR02018	Yeast Glc seven-like Phosphatase	*C. elegans*	NP_491237	$2e^{-94}$	no	184	13.0	2	4.74
Other functions										
518	SR00863	MFP2b	*A. suum*	AAP94889	$5e^{-71}$	no	173	52.0	7	14.85
519	SR00750	Similar to mannose receptor	*G. gallus*	XP_418617	$2e^{-07}$	yes	174	21.3	3	6.35
520	SR00671	Lysozyme family member (lys-5)	*C. elegans*	NP_502193	$4e^{-37}$	yes	160	23.1	4	12.70
521	SR00576	MSP domain protein	*B. malayi*	XP_001899679	$3e^{-32}$	no	97	47.4	4	8.02
522	SR01169	Aminotransferase	*C. botulinum*	ZP_02614737	0.53	no	176	31.8	4	8.14
523	SS00790	NompA	*C. elegans*	NP_502699	$5e^{-68}$	yes	160	12.5	2	4.00
524	SS01173	Enoyl-CoA reductase	*A. suum*	AAC48316	$1e^{-104}$	no	299	10.0	2	6.41
525	SR00354	Acid sphingomyelinase	*C. elegans*	NP_001040996	$2e^{-89}$	yes	269	19.3	3	6.09
526	SR02017	26 kDa secreted antigen	*T. canis*	P54190	$5e^{-04}$	yes	139	21.6	2	4.00
527	SR02511	Acyl sphingosine amino hydrolase	*C. briggsae*	CAP33700	$2e^{-48}$	yes	184	27.2	4	8.35
528	SS00929	High mobility group protein	*C. elegans*	NP_496970	$5e^{-21}$	no	94	20.2	2	4.02
529	SR00172	60s acidic ribosomal protein P2	*B. malayi*	P90703	$3e^{-21}$	no	110	27.3	2	4.00
530	SR00821	Saposin-like protein	*C. elegans*	NP_491803	5.4	yes	86	54.7	3	6.39
Not assigned										
531	SR00223	Hypothetical protein C50B6.7	*C. elegans*	NP_506303	$6e^{-51}$	yes	192	18.2	2	4.02
532	SR01917	Hypothetical protein F40F4.6	*C. elegans*	NP_508552	$6e^{-03}$	yes	166	18.7	2	4.00

Table 5a continued

	Cluster	BLAST Alignment	Species	Accession Number	E	SP	EST Lgt.	% Cov.	№ Pep.	UPS
533	SR00294	Hypothetical protein R05F9.12	C. elegans	AAA83174	$3e^{-49}$	yes	168	21.4	2	4.00
534	SR00380	Hypothetical protein EUBVEN_01944	E. ventriosum	ZP_02026680	7.9	yes	154	14.9	2	5.78
535	SR00282	Hypothetical protein F40F4.6	C. elegans	NP_508552	$2e^{-28}$	yes	178	15.7	2	4.00
536	SR00716	F09C8.1	C. elegans	NP_510636	$3e^{-45}$	yes	170	22.4	2	4.66
537	SR00375	Hypothetical protein CBG05949	C. briggsae	XP_001670383	$5e^{-09}$	yes	148	31.1	5	13.91
538	SR00767	F25A2.1	C. elegans	NP_503390	$6e^{-25}$	no	178	17.4	2	6.67
539	SR01936	Hypothetical protein CBG21853	C. briggsae	XP_001672742	$1e^{-21}$	yes	190	13.7	2	5.98
540	SR02994	Hypothetical protein Y49E10.18	C. elegans	NP_499623	$2e^{-27}$	yes	143	51.0	7	19.13
541	SR00899	Hypothetical protein CBG09313	C. briggsae	XP_001674244	$4e^{-11}$	no	231	13.9	2	4.01
542	SR02091	Hypothetical protein CBG22129	C. briggsae	XP_001667627	$1e^{-35}$	yes	174	36.8	6	13.27

9.1.10 Table 5b: Nematode RefSeq proteins only found in E/S products from the free-living stages

	Acc. Number	BLAST Alignment	Species	SP	Frag. Lgt.	% Cov.	№ Pep.	UPS
543	gi|1480463	Cyclophilin Dicyp-2	D. immitis	no	171	10.5	2	4.22
544	gi|17542416	Major sperm protein (msp-78)	C. elegans	no	127	25.2	2	4.70

9.1.11 Table 6a: List of *Strongyloides* EST cluster numbers found in E/S products from infective larvae and free-living stages

	Cluster	BLAST Alignment	Species	Accession Number	E	SP	EST Lgt.	% Cov.	№ Pep.	UPS
Protein digestion and folding										
545	SR01181	Aminopeptidase P, N-terminal domain	*B. malayi*	XP_00189318	$1e^{-45}$	no	172	18.6	3	6.85
546	SS01124	FKBP-type peptidyl-prolyl cis-trans isomerase	*B. malayi*	XP_00189823	$8e^{-107}$	yes	273	11.0	2	4.19
547	SS00794	Proteasome A- and B-type family protein	*B. malayi*	XP_00190249 8	$2e^{-51}$	no	163	12.3	2	4.01
548	SS02556	Aminopeptidase W07G4.4	*B. malayi*	XP_00190157 4	$2e^{-23}$	no	172	17.4	2	6.46
549	SS02960	Aspartyl aminopeptidase	*B. malayi*	XP_00189479 5	$4e^{-60}$	no	170	14.1	2	4.00
550	SR01240	Ube1-prov protein	*B. malayi*	XP_00190157 3	$4e^{-82}$	no	195	21.5	2	4.00
Carbohydrate metabolism										
551	SR04370	Phosphoglycerate kinase	*B. malayi*	XP_00189189 2	$4e^{-74}$	no	195	34.4	4	8.01
552	SR02292	Glycosyl transferase family 39	*A. chlorophenolicus*	ZP_0283933 3	0.18	no	172	16.3	2	4.00
Other functions										
553	SR00807	Nascent polypeptide-associated complex	*B. malayi*	XP_00189279 6	$6e^{-54}$	no	197	33.0	3	7.30
554	SS01433	Calmodulin	*C. elegans*	O16305	$1e^{-79}$	no	150	22.7	2	4.98
555	SR00518	Vacuolar ATP synthase catalytic subunit A	*B. malayi*	XP_00190166 0	$8e^{-125}$	no	257	19.5	3	7.39
556	SR00698	Cysteine synthase	*C. reinhardtii*	XP_00169193 5	$4e^{-54}$	no	165	27.9	3	7.53
557	SS02969	Aldo/keto reductase family protein	*B. malayi*	XP_00189774 3	$6e^{-36}$	no	138	26.1	2	5.52
558	SR00487	Vacuolar ATPase protein 8	*B. malayi*	XP_00190206 8	$1e^{-62}$	no	169	21.9	3	6.28
559	SS03304	Vacuolar ATP synthase subunit B	*B. malayi*	XP_00189387 2	$3e^{-53}$	no	115	30.4	2	4.00
560	SR00602	Probable s-adenosylmethionine synthetase	*C. elegans*	P50306	$3e^{-107}$	no	284	29.9	5	13.81
561	SR00810	Temporarily assigned gene name family member	*C. briggsae*	CAP37396	$2e^{-97}$	no	205	21.0	3	6.00
Not assigned										
562	SR00882	Unknown protein	*A. thaliana*	AAL32597	0.44	no	345	21.7	8	17.41
563	SR02501	Hypothetical protein	*Y. lipolytica*	XP_500483	0.42	no	179	19.0	3	10.27

9.1.12 Table 6b: Nematode RefSeq proteins found in E/S products from infective larvae and free-living stages

	Acc. Number	BLAST Alignment	Species	SP	Frag. Lgt.	% Cov.	№ Pep.	UPS
564	gi|17535051	Hypothetical protein K10H10.2	*C. elegans*	no	337	6.8	2	4.00

9.1.13 Table 7a: List of *Strongyloides* EST cluster numbers found in E/S products from parasitic females and free-living stages

	Cluster	BLAST Alignment	Species	Accession Number	E	SP	EST Lgt.	% Cov.	№ Pep.	UPS
Protein digestion and folding										
565	SR00486	Aspartyl protease 3	*C. briggsae*	CAP30177	$1e^{-67}$	yes	193	56.5	8	16.01
566	SR00053	Aspartyl protease precursor	*S. stercoralis*	AAD09345	0.0	trun	368	18.7	7	16.59
567	SR00889	Aspartyl protease 3	*C. briggsae*	CAP30177	$2e^{-68}$	yes	188	36.2	5	12.25
Other functions										
568	SR03308	Immunosuppressive ovarian message protein	*A. suum*	CAK18210	$1e^{-15}$	yes	240	7.5	2	4.00
569	SR01801	Ornithine aminotransferase	*C. elegans*	Q18040	$1e^{-80}$	no	202	20.8	3	6.56
570	SR00157	Lysozyme family member (lys-8)	*C. elegans*	NP_495083	$6e^{-20}$	yes	124	12.1	2	5.05
571	SR00961	ATP synthase alpha chain	*B. malayi*	XP_00190075	0.0	yes	364	14.3	4	8.21
572	SR00779	Probable voltage-dependent anion-selective channel	*C. elegans*	Q21752	$1e^{-42}$	no	238	32.0	5	11.60
573	SR00101	DAF-16/FOXO controlled, germ-line tumor affecting family member	*C. elegans*	NP_496755	$4e^{-18}$	yes	177	41.2	7	15.02
574	SR02116	Surface-associated antigen 2	*N. americanus*	ACE79378	$7e^{-30}$	yes	150	21.3	3	6.31
575	SR01068	Ribosomal protein (rpl-1)	*C. elegans*	NP_491061	$2e^{-88}$	no	216	12.0	2	4.11
576	SR00871	Lysozyme family member (lys-4)	*C. elegans*	NP_502192	$1e^{-53}$	yes	178	55.1	8	18.74
577	SS00016	Ribosomal protein (rpl-12)	*C. elegans*	NP_502542	$5e^{-78}$	no	166	15.1	2	4.91
578	SR00047	Acidic ribosomal protein	*S. ratti*	ABF69530	$8e^{-32}$	no	111	14.5	2	4.00
Not assigned										
579	SR02493	Hypothetical protein T25C12.3	*C. elegans*	NP509919	$5e^{-23}$	yes	177	23.7	2	4.94
580	SR00815	Hypothetical protein CBG22718	*C. briggsae*	XP_00167751 0	$7e^{-48}$	yes	320	18.4	7	17.14
581	SR00540	Hypothetical protein T25C12.3	*C. elegans*	NP_498952	1.6	no	182	24.2	3	6.03
582	SR00311	Hypothetical protein C24B9.3a	*C. elegans*	NP_00102367 1	$2e^{-10}$	yes	154	52.6	5	10.42
583	SR00874	Hypothetical protein	*P. tetraurelia*	XP_00143799 5	1.8	no	164	12.2	2	4.00
584	SR00773	Hypothetical protein F28B4.3	*C. elegans*	NP_508551	$3e^{-19}$	yes	395	7.3	2	4.57
585	SR00528	Hypothetical protein F28B4.3	*C. elegans*	XP_00167637 3	$6e^{-22}$	yes	168	17.9	2	4.01
586	SR00557	Hypothetical protein CBG06065	*C. briggsae*	XP_00167028 5	$5e^{-20}$	yes	158	24.7	3	8.83

9.1.14 Table 8a: List of *Strongyloides* EST cluster numbers found in extracts from the parasitic, the infective and the free-living stages

	Cluster	BLAST Alignment	Species	Accession Number	E	SP	EST Lgt.	% Cov.	№ Pep.	UPS
Oxydative metabolism										
587	SS03191	Succinate semialdehyde dehydrogenase	A. aegypti	XP_001649617	$1e^{-44}$	no	159	22.0	2	4.03
588	SR00908	Ubiquinol-Cytochrome c oxidoreductase complex	C. elegans	NP_498202	$7e^{-109}$	no	386	50.1	12	29.03
589	SR00586	Uncharacterised oxidoreductase W01C9.4	C. elegans	Q23116	$3e^{-38}$	no	167	34.7	3	7.77
590	SR01028	Cytochrome C family member	C. elegans	NP_492207	$1e^{-67}$	yes	199	43.2	5	12.17
591	SR00914	Iron-sulfur protein	C. briggsae	CAP23049	$2e^{-98}$	no	264	6.4	2	8.95
592	SR00164	Cytochrome C oxidase	C. briggsae	CAP33937	$6e^{-61}$	no	163	21.5	3	9.72
Carbohydrate Metabolism										
593	SS02197	Acetyl-CoA hydrolase/transferase	B. malayi	XP_001900057	$1e^{-60}$	no	167	35.3	3	6.00
Cytosol energy metabolism										
594	SR03331	Propionyl CoA carboxylase alpha subunit	C. elegans	NP_509293	$5e^{-75}$	no	262	56.1	12	30.03
595	SR00726	Propionyl CoA carboxylase beta subunit	C. elegans	NP_741742	$6e^{-137}$	no	276	49.6	9	19.67
596	SS01672	Enoyl-CoA hydratase	C. elegans	NP_506810	$6e^{-153}$	no	464	20.5	5	10.95
597	SR02350	Acyl-CoA dehydrogenase	B. malayi	XP_001896268	$2e^{-61}$	no	180	25.0	4	9.70
598	SS00570	Acyl-CoA dehydrogenase	B. malayi	XP_001896268	$8e^{-73}$	no	212	34.4	5	10.00
599	SR03204	3-Ketoacyl-CoA thiolase	C. briggsae	CAP21279	$1e^{-56}$	no	158	34.8	3	8.08
Protein biosynthesis										
600	SS02711	Heat-responsive protein 12	R. norvegicus	NP_113902	$6e^{-19}$	no	66	50.0	3	6.00
Protein digestion and folding										
601	SR00755	Peptidase M16 inactive domain containing protein	B. malayi	XP_001898901	$3e^{-72}$	no	289	31.1	5	10.54
602	SR00897	Aspartyl protease	C. briggsae	CAP25783	$3e^{-58}$	yes	189	20.1	3	6.01
Structural proteins										
603	SS00331	Myosin heavy chain	T. spiralis	AAK54395	$2e^{-43}$	trun	148	35.1	5	11.70
Nucleic acid metabolism										
604	SR01030	40S ribosomal protein S5	B. malayi	XP_001897018	$2e^{-99}$	no	208	51.0	6	15.34
605	SR01020	40S ribosomal protein S7	B. malayi	XP_001893453	$2e^{-62}$	no	191	29.3	5	13.77

Table 8a continued

	Cluster	BLAST Alignment	Species	Accession Number	E	SP	EST Lgt.	% Cov.	№ Pep.	UPS
606	SR00965	60S ribosomal protein L19	B. malayi	XP_001898252	$4e^{-76}$	no	199	19.6	4	10.85
607	SR01038	60S ribosomal protein L18a	B. malayi	XP_001896063	$1e^{-68}$	no	172	40.7	8	16.96
608	SS01318	60S ribosomal protein L17	B. malayi	XP_001892797	$4e^{-66}$	no	168	23.8	4	10.19
609	SR00804	40S ribosomal protein S11	B. malayi	XP_001898199	$3e^{-58}$	no	157	29.3	4	10.14
610	SS01522	Ribosomal protein 1	L. obliqua	AAV91380	$2e^{-35}$	no	158	26.0	3	6.08
611	SR00970	Ribosomal protein, large subunit	C. elegans	NP_494932	$9e^{-35}$	no	134	28.4	5	11.71
612	SS01392	Ribosomal protein S14	S. ratti	ABF69533	$1e^{-82}$	no	151	45.0	5	10.00
613	SR00964	40S Ribosomal protein S16	B. malayi	XP_001894284	$2e^{-53}$	no	146	39.0	7	18.11
614	SR00811	Ribosomal protein, small subunit	C. elegans	NP_740944	$6e^{-43}$	no	120	39.2	3	7.46
615	SS01083	40S Ribosomal protein S17	B. malayi	XP_001894318	$3e^{-50}$	no	127	44.1	6	14.31
616	SS01552	Ribosomal protein, small subunit	C. elegans	NP_502365	$8e^{-68}$	no	143	28.0	2	6.65
617	SR01005	60S Ribosomal protein L30	B. malayi	XP_001898088	$3e^{-49}$	no	112	52.7	6	13.80
618	SR00827	40S Ribosomal protein S15a	B. malayi	XP_001901171	$4e^{-60}$	no	133	39.1	4	11.77
619	SR00926	60S Ribosomal protein L23	B. malayi	XP_001900778	$1e^{-65}$	yes	139	46.0	5	11.23
620	SS01509	Ribosomal protein, small subunit	C. elegans	NP_493571	$1e^{-48}$	no	117	31.6	3	12.02
621	SS03372	Small nuclear ribonucleoprotein	S. bicolor	ACE86405	$1e^{-37}$	no	116	33.6	3	7.63
Cytoplasmatic										
622	SR00106	Threonyl-tRNA synthetase	B. malayi	XP_001893343	$2e^{-79}$	no	175	22.3	3	6.10
Other functions										
623	SR00538	Flavoprotein subunit of complex II	C. elegans	BAA21637	$4e^{-85}$	no	169	45.0	5	11.10
624	SR00416	ATP synthase subunit alpha	C. elegans	Q9XXK1	$4e^{-32}$	no	89	54.0	3	9.07
625	SR02856	Calsequestrin family protein	B. malayi	XP_001899493	$2e^{-51}$	yes	133	46.6	4	8.00
626	SS00622	DNA-binding protein	B. malayi	XP_001894006	$2e^{-71}$	no	284	21.5	5	11.73
627	SR00927	Mitochondrial prohibitin complex protein 2	C. elegans	P50093	$1e^{-120}$	yes	277	19.1	5	10.01
628	SR00607	ATP Synthase B homolog family member	C. briggsae	XP_001676578	$1e^{-87}$	no	243	19.3	5	10.14
629	SR00928	ATP synthase subunit	C. elegans	NP_001021420	$7e^{-64}$	no	205	58.5	9	20.20
630	SR00717	ATP synthase subunit f	C. elegans	Q22021	$9e^{-60}$	no	152	21.7	3	8.10

Table 8a continued

	Cluster	BLAST Alignment	Species	Accession Number	E	SP	EST Lgt.	% Cov.	№ Pep.	UPS
631	SR00705	ATP Synthase G homolog	*C. elegans*	NP_492352	$2e^{-42}$	no	130	32.4	4	8.00
632	SR00280	RNAi-induced longevity family member	*C. briggsae*	CAP25262	$7e^{-38}$	no	128	54.7	6	12.01
Not assigned										
633	SR02752	Hypothetical protein Y69A2AR.18a	*C. elegans*	AAK68562	$3e^{-71}$	yes	206	22.8	4	9.52
634	SR00959	Hypothetical protein CBG19690	*C. briggsae*	XP_001671790	$2e^{-51}$	no	139	41.7	4	8.01
635	SS01396	Hypothetical protein CBG09544	*C. briggsae*	XP_001674053	$5e^{-66}$	no	193	16.6	3	6.01
636	SS02192	Aly	*D. yakuba*	XP_002096710	$4e^{-11}$	no	168	25.6	4	8.14
637	SR00643	Hypothetical protein CBG23568	*C. briggsae*	CAP20392	$8e^{-43}$	no	137	48.2	5	10.00
638	SR02885	Hypothetical protein W09C5.8	*C. elegans*	NP_493394	$2e^{-47}$	no	156	39.1	5	12.66
639	SS00764	Sr-mps-1 protein	*S. ratti*	CAA04549	$5e^{-43}$	no	84	38.1	3	6.97

9.1.15 Table 8b: Nematode RefSeq proteins found in extracts from the parasitic, the infective and the free-living stages

	Acc. Number	BLAST Alignment	Species	SP	Frag. Lgt.	% Cov.	№ Pep.	UPS
640	gi|17554946	Enoyl-CoA hydratase family member	*C. elegans*	no	288	9.7	2	4.58

9.1.16 Table 9a: List of *Strongyloides* EST cluster numbers found only in extracts from infective larvae

Cluster	BLAST Alignment	Species	Accession Number	E	SP	EST Lgt.	% Cov.	№ Pep.	UPS
Oxydative metabolism									
641 SR03791	NADH ubiquinone oxidoreductase	C. elegans	NP_872121	$5e^{-59}$	no	190	14.2	2	5.92
642 SR00207	FeS assembly protein SufB	S. pneumoniae	ZP_01827340	0.61	no	132	24.2	3	9.13
643 SR00500	Aldo/keto reductase	B. malayi	XP_00189773	$2e^{-48}$	no	172	21.5	2	4.92
644 SR01377	GILT-like protein	C. elegans	O17861	$6e^{-30}$	yes	180	17.8	4	9.57
645 SR01129	NADH-ubiquinone oxidoreductase	C. elegans	Q9N2W7	$2e^{-42}$	no	132	22.7	2	4.04
646 SR02324	Cytochrome B large subunit	A. suum	BAA11232	$6e^{-39}$	no	168	25.6	3	6.01
647 SS00100	NADH dehydrogenase	C. elegans	Q18359	$2e^{-47}$	no	154	24.0	2	4.00
Carbohydrate Metabolism									
648 SS02410	Phosphofructokinase	B. malayi	XP_00189768	$4e^{-65}$	no	214	12.1	2	4.57
649 SS02884	Phosphofructokinase	A. suum	AAR16088	$2e^{-60}$	no	185	18.4	2	4.01
Cytosol energy metabolism									
650 SS02205	Fatty acid CoA synthetase	C. elegans	NP_508993	$7e^{-91}$	no	201	20.4	3	6.00
651 SR02385	Enoyl-CoA hydratase	C. elegans	NP_506810	$3e^{-70}$	no	189	10.6	2	4.00
Structural proteins									
652 SS02656	Kettin 1	C. briggsae	CAP26642	$2e^{-65}$	trun	203	33.0	4	8.27
653 SS00744	Alpha collagen	C. elegans	CAA40299	$3e^{-90}$	yes	187	31.0	3	6.00
654 SR00067	Alpha tubulin	M. mulatta	XP_001108104	$1e^{-58}$	no	187	37.4	4	7.10
Other functions									
655 SS00549	Propionyl-coenzyme A carboxylase	C. brenneri	ACD88885	$3e^{-48}$	yes	152	20.4	3	7.98
656 SR04479	Vacuolar H ATPase	C. elegans	NP_508711	$2e^{-65}$	no	154	17.5	2	5.71
657 SS01310	G-o protein	C. elegans	AAA28059	$2e^{-83}$	no	165	40.0	6	13.47
658 SS03197	Aquaglyceroporin	C. elegans	NP_508515	$2e^{-34}$	no	219	12.8	2	4.00
659 SR00079	Histone 1	C. familiaris	XP853903	$6e^{-73}$	no	148	21.6	2	5.26
660 SS01339	Vacuolar ATP synthase	C. briggsae	Q612A5	$2e^{-80}$	no	161	38.5	3	6.23
Not assigned									
661 SR04691	Hypothetical protein ZC376.2	C. elegans	NP_506509	$1e^{-19}$	yes	154	13.0	2	4.01
662 SR03184	Hypothetical protein B0303.3	C. elegans	NP_498915	$3e^{-68}$	no	187	15.5	2	4.22
663 SR00405	Hypothetical protein CBG09503	C. briggsae	XP_00167490	$2e^{-53}$	no	150	14.7	2	4.19
664 SS00166	1810034M08Rik protein	B. malayi	XP_00189623	$2e^{-55}$	no	171	23.4	3	6.00

9.1.17 Table 9b: Nematode RefSeq proteins found only in extracts from infective larvae

	Acc. Number	BLAST Alignment	Species	SP	Frag. Lgt.	% Cov.	№ Pep.	UPS
665	gi|71998537	Uncoordinated family member 52	C. elegans	yes	3,375	1.0	3	6.83
666	gi|71999957	Uncoordinated family member 70	C. elegans	no	2,302	1.0	2	5.23
667	gi|603211	Glyceraldehyde 3-phosphate dehydrogenase	B. malayi	no	339	13.0	2	5.92
668	gi|71993575	Hypothetical protein K09E2.2	C. elegans	no	552	6.3	2	4.10
669	gi|17539608	Egg Laying defective family member (egl-4)	C. elegans	no	780	3.3	2	6.58
670	gi|17531727	NADH Ubiquinone oxidoreductase	C. elegans	no	479	8.8	2	7.45
671	gi|784942	Cytoplasmic intermediate filament protein	A. lumbricoides	no	612	2.9	2	4.15
672	gi|17544026	Hypothetical protein Y69A2AR.18a	C. elegans	yes	299	9.0	2	4.10
673	gi|71982447	RAB family member	C. elegans	no	233	16.3	3	6.35
674	gi|71999796	Lethal family member 60	C. elegans	no	184	16.3	2	4.85

9.1.18 Table 10a: List of *Strongyloides* EST cluster numbers found only in extracts from parasitic females

Cluster		BLAST Alignment	Species	Accession Number	E	SP	EST Lgt.	% Cov.	№ Pep.	UPS
Oxydative metabolism										
675	SR00505	Dihydro-pyrimidine dehydrogenase	*M. musculus*	Q8CHR6	$4e^{-69}$	no	185	15.1	2	4.86
676	SS00714	Phosphoenol-pyruvate carboxykinase	*H. contortus*	P29190	$1e^{-52}$	no	139	27.3	2	4.47
677	SR02382	Glucocerebrosidase	*A. pisum*	XP_001945612	$1e^{-11}$	no	82	42.7	2	4.00
678	SR03739	Glyceraldehyde-3-phosphate dehydrogenase	*R. norvegicus*	XP_237427	$3e^{-39}$	no	94	14.9	2	4.46
Cytosol energy metabolism										
679	SS01093	Fatty acid synthase	*C. briggsae*	CAP31392	$8e^{-123}$	no	279	7.9	2	4.31
680	SS00238	Enoyl-CoA hydratase	*C. elegans*	NP_499156	$5e^{-52}$	no	127	22.8	2	4.30
Protein metabolism										
681	SR01222	H/ACA ribonucleoprotein complex subunit 4	*C. briggsae*	Q60YA8	$4e^{-74}$	no	174	35.6	4	8.08
682	SS02122	Protein synthesis factor	*B. malayi*	XP_001898787	$2e^{-52}$	no	165	16.4	2	6.17
Protein digestion and folding										
683	SR04440	Prolyl-oligopeptidase	*S. ratti*	ACH87625	$2e^{-28}$	yes	191	11.5	2	7.15
684	SR01697	Aminopeptidase 1	*C. elegans*	AAC70927	$1e^{-35}$	no	207	15.0	2	9.06
685	SR00655	26S proteasome subunit	*S. salar*	ACH85272	$6e^{-66}$	trun	233	15.5	2	4.00
686	SS02537	Cathepsin Z family member	*C. elegans*	NP_491023	$2e^{-18}$	yes	51	51.0	2	4.00
Structural proteins										
687	SR01204	Clathrin light chain family protein	*B. malayi*	XP_001902771	$1e^{-18}$	no	227	12.8	2	4.02
688	SS00300	Myosin light chain family member	*C. elegans*	NP_497700	$9e^{-70}$	no	161	18.0	2	5.29
689	SS01088	Troponin C family member	*C. elegans*	NP_496251	$1e^{-67}$	no	159	11.3	2	5.15
Heat-shock proteins										
690	SS02424	DnaK protein	*B. malayi*	XP_001899323	$1e^{-17}$	no	188	33.5	3	6.82
691	SR00126	Hsp90 co-chaperone Cdc37	*B. malayi*	XP_001898527	$4e^{-46}$	no	286	21.7	4	8.03
692	SR00156	Tubulin-specific chaperone B	*C. elegans*	Q20728	$2e^{-21}$	no	191	28.8	4	10.93
693	SR05119	Small heat-shock protein	*B. malayi*	XP_001895416	$1e^{-23}$	trun	155	20.0	3	6.20
Cytoplasmatic										
694	SS01105	Isoleucyl tRNA Synthetase	*C. elegans*	NP_501914	$1e^{-92}$	no	314	11.1	4	10.52

Table 10a continued

	Cluster	BLAST Alignment	Species	Accession Number	E	SP	EST Lgt.	% Cov.	№ Pep.	UPS
695	SR02648	Leucyl tRNA Synthetase	C. elegans	NP_497837	$7e^{-73}$	no	216	10.6	2	4.40
696	SR02316	Valyl-tRNA synthetase	B. malayi	XP_00189806	$3e^{-53}$	no	173	15.6	2	4.94
697	SR04833	Valyl-tRNA synthetase	C. elegans	NP_493377	$2e^{-66}$	no	187	10.7	2	4.25
698	SS02671	Alanyl tRNA synthetase	B. malayi	NP_491281	$4e^{-50}$	no	179	21.8	2	4.00
699	SR01288	Glutamyl tRNA synthetase	C. elegans	NP_492711	$4e^{-120}$	no	253	9.9	2	4.75
700	SR00796	Eukaryotic translation initiation factor	B. malayi	XP_001893193	$1e^{-19}$	yes	163	19.0	2	4.64
701	SR01969	Signal recognition particle	B. malayi	XP_001899803	$7e^{-28}$	no	146	26.0	2	5.52
Nucleic acid metabolism										
702	SR00823	60S ribosomal protein L35	B. malayi	XP_001898017	$2e^{-38}$	no	120	17.5	2	4.26
703	SR00713	Mitochondrial single stranded DNA binding protein	C. elegans	NP_498935	$1e^{-10}$	no	176	13.1	2	6.57
704	SR02658	39S ribosomal protein L23	C. elegans	Q9GYS9	$2e^{-40}$	no	147	13.6	2	4.56
705	SS00141	DNA-directed RNA polymerase II	B. malayi	XP_001896469	$5e^{-38}$	no	121	24.0	2	4.00
706	SR00886	Ribosomal protein L34	S. purpuratus	XP_797232	$8e^{-34}$	no	111	22.5	2	4.48
707	SR00958	60S ribosomal protein L36	B. malayi	XP_001894797	$2e^{-32}$	no	105	35.2	4	11.26
708	SR01016	Ribosomal protein S25	B. elongata	ABW90430	$2e^{-29}$	no	115	21.7	2	4.13
709	SS01085	Ribosomal small subunit protein 30	C. remanei	AAY46303	$1e^{-26}$	trun	105	11.4	2	4.00
Other functions										
710	SR03671	Chitin-binding peritrophin-A domain containing protein	B. malayi	XP_001895704	$2e^{-15}$	no	195	14.9	2	5.40
711	SR03049	Trichohyalin	T. vaginalis	XP_001327090	$5e^{-05}$	no	103	18.4	2	4.51
712	SS01271	CUT domain containing protein	B. malayi	XP_001900236	$2e^{-50}$	no	222	15.8	2	4.04
713	SR01023	ALT-1	S. ratti	AAT79348	0.0	no	331	17.2	3	14.75
714	SS02067	Gamma-glutamyl transpeptidase	B. malayi	AAD09400	$2e^{-47}$	yes	155	18.1	2	4.11
715	SR04742	Cytokinesis, apoptosis, RNA-associated family member	C. elegans	NP_493254	$1e^{-38}$	no	126	24.6	2	4.00
716	SS00841	Transformer-2a3	B. malayi	XP_001900358	$1e^{-30}$	no	162	15.4	2	6.13
717	SR03312	Surf2	B. malayi	XP_001893176	$1e^{-14}$	no	160	14.4	2	5.38
718	SR00432	H1 histone isoform H1.1	C. remanei	AAV50096	$4e^{-12}$	trun	113	24.8	2	5.31
719	SR01125	Proliferating cell nuclear antigen	C. briggsae	CAP32191	$3e^{-78}$	no	251	22.7	4	8.22

Table 10a continued

	Cluster	BLAST Alignment	Species	Accession Number	E	SP	EST Lgt.	% Cov.	№ Pep.	UPS
720	SR00988	Histone H1-delta	S. purpuratus	NP_999722	$8e^{-18}$	no	160	14.4	2	6.93
721	SR02156	Vacuolar H ATPase	C. remanei	ABW36063	$6e^{-08}$	trun	139	32.4	2	4.34
722	SR00170	Probable ATP-dependent RNA helicase ddx20	D. discoideum	XP_642653	0.02	no	177	12.4	2	4.26
723	SS00776	Probable vacuolar proton pump subunit H 2	C. briggsae	Q619W9	$4e^{-99}$	no	255	11.0	2	4.03
724	SR01142	Lipid depleted family member	C. elegans	NP_491359	$1e^{-52}$	no	178	15.7	3	6.21
725	SR03278	FIS 1 related protein	B. malayi	XP_001896808	$4e^{-34}$	no	178	16.3	2	4.00
Not assigned										
726	SR03717	Similar to CG3108-PA	R. norvegicus	XP_00106632 6	3.2	no	194	58.2	8	16.07
727	SR00210	Hypothetical protein	B. malayi	XP_001897658	$7e^{-17}$	no	166	21.7	3	6.42
728	SR01616	Hypothetical protein K10C2.4	C. elegans	NP_509083	$9e^{-55}$	no	166	10.2	2	4.46
729	SR03861	Hypothetical protein F28A10.6	C. elegans	NP_493832	$1e^{-69}$	no	191	18.8	2	4.01
730	SR00612	Hypothetical protein RE06140p	B. malayi	XP_001894655	$1e^{-45}$	no	240	18.7	2	4.00
731	SS00932	Hypothetical protein GK13574	D. willistoni	XP_002072650	$2e^{-17}$	no	167	15.0	2	4.02
732	SR01570	Hypothetical protein	B. malayi	XP_001900350	$4e^{-61}$	no	150	17.3	2	4.00
733	SR05160	Putative retroelement	O. sativa	AAN04209	0.88	no	191	10.5	2	5.23
734	SS01526	RIKEN cDNA 2610002M06	B. malayi	XP_001896640	$1e^{-75}$	trun	206	12.1	2	4.94
735	SR01941	Hypothetical protein F23F1.2	C. elegans	NP_493641	$4e^{-21}$	yes	181	23.2	2	7.40
736	SR00754	Hypothetical protein Bm1_37315	B. malayi	XP_001898912	$2e^{-04}$	no	158	13.9	2	5.55
737	SR00589	hypothetical protein CBG18325	C. briggsae	XP_001665919	$1e^{-40}$	no	172	16.3	2	6.17
738	SR02260	hypothetical protein Y43F8C.8	C. elegans	NP_507808	$6e^{-29}$	no	160	15.0	2	5.30

9.1.19 Table 10b: Nematode RefSeq proteins found only in extracts from parasitic females

	Acc. Number	BLAST Alignment	Species	SP	Frag. Lgt.	% Cov.	№ Pep.	UPS
739	gi|17540638	Hypothetical protein F55B11.1	C. elegans	no	1,358	1.4	2	7.15
740	gi|71986512	Hypothetical protein F14E5.2b	C. elegans	yes	1,147	3.2	3	6.25
741	gi|17506981	Alanyl tRNA synthetase family member	C. elegans	no	968	2.4	3	7.32
742	gi|25144732	Maternal effect lethal family member	C. elegans	yes	507	5.7	2	8.01
743	gi|17508577	Hypothetical protein R06C7.5a	C. elegans	no	478	8.7	2	5.16
744	gi|17553978	Hypothetical protein K01G5.5	C. elegans	no	445	4.0	2	5.13
745	gi|17539424	BT toxin resistant family member	C. elegans	no	399	4.8	2	4.41
746	gi|25153153	Hypothetical protein W03F8.1	C. elegans	no	194	8.8	2	4.97
747	gi|32566061	CalciNeurin B family member 1	C. elegans	no	171	23.4	3	7.05

9.1.20 Table 11a: List of *Strongyloides* EST cluster numbers found in extracts from infective larvae and parasitic females

Cluster		BLAST Alignment	Species	Accession Number	E	SP	EST Lgt.	% Cov.	№ Pep.	UPS
Oxydative metabolism										
748	SS02280	2-oxoglutarate dehydrogenase	C. elegans	O61199	$4e^{-86}$	no	218	21.6	4	8.04
749	SR00722	NADH-ubiquinone oxidoreductase	B. malayi	XP_00190 18 94	$2e^{-72}$	no	279	12.9	3	9.34
750	SR02139	Flavoprotein subunit of succinate dehydrogenase	A. suum	BAB84191	$1e^{-70}$	no	180	23.3	4	9.84
751	SR03424	Electron transfer flavoprotein subunit alpha	C. elegans	Q93615	$3e^{-72}$	no	240	20.0	3	6.26
Cytosol energy metabolism										
752	SR03196	Acetyl CoA acyl-transferase	B. taurus	NP001030419	$7e^{-68}$	no	208	14.9	2	9.84
Protein digestion and folding										
753	SS02828	Proteasome A-type and B-type family protein	B. malayi	XP_00190 1796	$8e^{-08}$	no	83	62.7	4	10.00
754	SS00868	Proteasome A-type and B-type family protein	B. malayi	XP_00190 1133	$2e^{-48}$	no	192	39.1	3	7.40
Structural proteins										
755	SS02178	Intermediate filament protein	A. lumbricoides	CAA60046	$4e^{-72}$	yes	163	21.4	3	10.64
756	SR00472	Intermediate filament protein	C. briggsae	CAP35058	$4e^{-68}$	no	189	9.5	2	4.47
757	SR01304	Lamin	P. caudatus	CAB43347	$7e^{-11}$	no	163	17.8	3	7.26
Developmental processes										
758	SR00869	LET-721 protein	C. briggsae	CAP36880	$7e^{-72}$	no	191	24.1	2	6.81
Heat-shock proteins										
759	SS01424	Heat-shock protein 25	C. elegans	NP_001024375	$3e^{-75}$	no	211	40.3	6	14.94
Cytoplasmatic										
760	SS01404	Glutamine synthetase	C. briggsae	CAP32547	$1e^{-175}$	no	391	6.4	2	7.16
Other functions										
761	SR00762	Lin-5 interacting protein	C. briggsae	CAP33058	$9e^{-63}$	no	292	16.4	4	10.39
762	SR00781	Plant late embryo abundant related family member	C. elegans	NP_00102 4042	$4e^{-04}$	no	282	30.1	8	25.06
763	SS01277	Endoplasmin precursor	B. malayi	XP_00189 9398	$8e^{-93}$	yes	385	11.9	3	6.07
764	SS00250	Methylmalonyl CoA mutase	C. elegans	NP_497786	$2e^{-56}$	no	163	26.4	3	6.92

Table 11a continued

	Cluster	BLAST Alignment	Species	Accession Number	E	SP	EST Lgt.	% Cov.	№ Pep.	UPS
765	SR02509	TRIM5/cyclophilin A V3 fusion protein	*A. trivirgatus*	AAT99908	1.3	no	120	30.8	3	6.72
766	SS01130	ATP synthase subunit delta	*C. elegans*	Q09544	$1e^{-46}$	no	167	26.9	2	4.27
Not assigned										
767	SR00603	Hypothetical protein CBG07713	*C. briggsae*	XP_001677715	$6e^{-28}$	no	157	24.2	2	4.01
768	SR00631	Hypothetical protein CBG19014	*C. briggsae*	XP_001674408	$7e^{-105}$	no	238	23.1	4	13.46
769	SR00808	Hypothetical protein F59B1.2	*C. elegans*	NP_504020	$1e^{-03}$	yes	191	16.8	4	9.54
770	SR00073	Hypothetical protein Y43F8B.1a	*C. elegans*	NP_507798	$2e^{-16}$	no	160	40.0	5	13.12
771	SR00793	Hypothetical protein CBG01794	*C. briggsae*	XP_001671142	$1e^{-37}$	no	140	15.7	2	5.30

9.1.21 Table 11b: Nematode RefSeq proteins found in extracts from infective larvae and parasitic females

	Acc. Number	BLAST Alignment	Species	SP	Frag. Lgt.	% Cov.	№ Pep.	UPS
772	gi|17567343	Propionyl CoA carboxylase alpha subunit	*C. elegans*	no	724	10.6	6	12.13

9.1.22 Table 12a: List of *Strongyloides* EST cluster numbers found only in extracts from the free-living stages

	Cluster	BLAST Alignment	Species	Accession Number	E	SP	EST Lgt.	% Cov.	№ Pep.	UPS
Oxydative metabolism										
773	SR00136	PQQ enzyme repeat family protein	*B. malayi*	XP_00190083	$1e^{-32}$	yes	275	32	6	13.40
774	SS02316	PQQ enzyme repeat family protein	*B. malayi*	XP_00190083	$1e^{-24}$	yes	202	17.8	3	7.41
775	SR03630	PQQ enzyme repeat family protein	*B. malayi*	XP_00190083	$2e^{-23}$	yes	191	13.1	2	4.31
776	SR04810	Aldehyde dehydrogenase	*S. sviceus*	YP_00220868	$9e^{-36}$	no	165	33.3	3	6.02
777	SS01307	Short chain dehydrogenase	*C. elegans*	NP_509146	$1e^{-79}$	no	183	39.3	4	11.16
778	SR02519	Coenzyme Q biosynthesis family member	*C. elegans*	NP_505415	$9e^{-40}$	no	188	13.3	2	4.00
779	SR03324	NADH ubiquinone oxidoreductase	*C. elegans*	NP_741215	$2e^{-78}$	yes	244	32.8	7	14.01
780	SS03159	Oxidoreductase, zinc-binding dehydrogenase	*B. malayi*	XP_00189385	$1e^{-34}$	no	175	12.6	2	4.38
781	SR02076	Short chain dehydrogenase	*C. elegans*	NP_00112508	$4e^{-56}$	no	188	35.6	4	8.58
782	SR02992	Glutathione synthetase	*C. elegans*	XP_00189253	$1e^{-29}$	no	181	9.9	2	4.14
783	SR00144	2-oxoisovalerate dehydrogenase beta subunit	*C. elegans*	XP_00190275	$1e^{-47}$	no	138	21.7	2	4.01
784	SR01193	Short chain dehydrogenase	*C. elegans*	NP_492563	$2e^{-36}$	no	181	13.8	2	4.00
785	SR04350	Glutathione synthetase	*B. malayi*	XP_00189253	$6e^{-22}$	no	170	10.6	2	5.61
786	SR01749	Succinate-ubiquinone reductase	*A. suum*	BAB84192	$1e^{-25}$	no	152	23.6	2	4.03
Carbohydrate metabolism										
787	SR02172	1 beta1,3-galactosyltransferase	*C. elegans*	Q18515	$1e^{-43}$	yes	184	20.7	3	6.00
788	SR03674	Glycogen synthase kinase 3	*D. japonica*	BAD93244	$8e^{-75}$	no	191	17.8	2	5.73
789	SR00758	Ribose phosphate diphosphokinase	*B. malayi*	XP_00189420	$2e^{-90}$	trun	192	22.4	3	6.00
790	SR04582	Phosphomannomutase 2	*B. malayi*	XP_00189914	$6e^{-39}$	no	167	27.5	4	8.06
Cytosol energy metabolism										
791	SR01174	Hydroxlacyl-Co A dehydrogenase	*A. proteobacterium BAL199*	ZP_02187294	1.6	yes	184	23.9	3	7.15
792	SR01901	Fatty acid CoA synthetase	*C. elegans*	NP_508993	$5e^{-76}$	no	198	20.7	3	6.44
793	SS01084	Fatty acid CoA synthetase	*C. briggsae*	CAP33209	$3e^{-66}$	no	193	11.4	2	4.07

Table 12a continued

	Cluster	BLAST Alignment	Species	Accession Number	E	SP	EST Lgt.	% Cov.	№ Pep.	UPS
794	SR02202	Long-chain-fatty-acid-CoA ligase	R. etli	YP_001977177	$3e^{-58}$	no	186	11.8	2	4.13
795	SR04357	Isocitrate dehydrogenase gamma subunit	S. stercoralis	AAD55084	$9e^{-103}$	no	205	9.3	2	4.48
796	SR03152	Acetyl-coenzyme A synthetase 2	B. malayi	XP_00190093	$2e^{-53}$	no	192	23.4	3	6.01
797	SR00110	Isocitrate dehydrogenase subunit beta	C. elegans	Q93353	$2e^{-14}$	no	145	35.2	3	6.01
798	SR00152	Isocitrate dehydrogenase subunit beta	C. elegans	Q93353	$7e^{-74}$	no	198	14.6	2	5.45
799	SS02223	Hydroxlacyl-Co A dehydrogenase	C. elegans	P41938	$8e^{-46}$	no	184	22.3	2	4.00
Protein metabolism										
800	SR01690	Initiation factor 5b	C. elegans	AAK68893	$2e^{-70}$	no	180	21.1	2	4.62
801	SR00475	Eukaryotic initiation factor 3s9	B. malayi	XP_00190252	$2e^{-19}$	trun	164	52.4	6	12.49
802	SR02881	Eukaryotic initiation factor 3s9	B. malayi	XP_00190252	$2e^{-28}$	trun	182	24.2	4	8.02
803	SR03260	Ribophorin II	S. purpuratus	XP_783899	$8e^{-11}$	yes	169	56.8	7	15.45
804	SR00883	Eukaryotic release factor	X. laevis	P35615	$3e^{-24}$	no	86	60.5	3	6.00
805	SR01422	Zinc finger transcription factor	C. elegans	NP_505526	$2e^{-85}$	no	201	14.9	2	5.10
806	SR02035	Histidyl tRNA synthetase	C. briggsae	CAP35319	$4e^{-67}$	no	190	10.0	2	4.01
807	SR03144	Lysyl tRNA synthetase	B. malayi	XP_001894758	$1e^{-109}$	no	225	21.3	4	11.37
808	SR00503	Eukaryotic initiation factor	C. elegans	NP_492785	$5e^{-89}$	no	247	13.4	2	9.19
809	SR00741	Eukaryotic initiation factor	C. briggsae	CAP27081	$2e^{-63}$	no	290	18.6	3	6.12
810	SS01814	Eukaryotic initiation factor 3 subunit 7	B. malayi	XP_001895380	$9e^{-64}$	no	211	11.4	2	4.26
811	SS00548	Eukaryotic initiation factor	C. elegans	NP_492785	$1e^{-40}$	no	153	20.3	2	4.00
812	SR00225	Eukaryotic translation initiation factor 3	B. mori	NP_00104033	$2e^{-07}$	no	84	29.8	2	4.05
813	SR00817	Protein kinase domain containing protein	B. malayi	XP_00189430	$7e^{-66}$	no	154	27.9	4	10.08
814	SR01132	Protein kinase domain containing protein	B. malayi	XP_00189430	$1e^{-68}$	no	155	31.0	4	8.01
815	SS00931	Eukaryotic initiation factor 3	C. elegans	NP_495988	$1e^{-27}$	no	238	20.2	3	7.62
816	SR00844	Aspartyl-tRNA synthetase	B. malayi	XP_00189703	$9e^{-124}$	no	338	7.1	2	4.41
817	SR02051	tRNA binding domain containing protein	T. gondii	EEB04979	$4e^{-55}$	trun	181	29.8	5	10.76

Table 12a continued

	Cluster	BLAST Alignment	Species	Accession Number	E	SP	EST Lgt.	% Cov.	№ Pep.	UPS
818	SR00799	Eukaryotic translation initiation factor 3	B. malayi	XP_00190259	$3e^{-48}$	no	163	23.9	3	6.82
819	SR02755	GTP-binding protein SAR1	B. malayi	XP_00189811	$5e^{-93}$	no	194	29.9	5	11.60
820	SS00207	Phenylalanyl-tRNA synthetase	C. elegans	Q19713	$2e^{-34}$	no	149	27.5	3	6.05
821	SR02291	Tyrosyl-tRNA synthetase	C. elegans	BAC76736	$1e^{-54}$	no	165	17.6	2	4.00
822	SS01512	Ubiquitin	B. malayi	AAL91108	$4e^{-39}$	no	162	13.6	2	4.00
Protein digestion and folding										
823	SR00295	26S Proteasome regulatory complex	B. malayi	XP_00189761	$1e^{-54}$	no	168	17.3	2	4.62
824	SR00720	Proteasome/ cyclosome repeat family protein	B. malayi	XP_00189799	$2e^{-45}$	no	167	21.6	3	6.00
825	SR01709	26S Proteasome regulatory complex	B. malayi	XP_00189761	$2e^{-55}$	no	179	19.0	2	4.00
826	SS00211	Proteasome regulatory particle	B. malayi	NP_498346	$5e^{-53}$	no	165	26.7	3	6.00
827	SR01461	Ubiquitin carboxyl terminal hydrolase	B. malayi	XP_001902802	$1e^{-26}$	no	195	30.3	4	9.47
828	SS00056	Calpain family protein 1	B. malayi	XP_00189560	$3e^{-57}$	no	153	18.3	2	4.47
829	SR02765	Putative serine protease	C. elegans	P90893	$2e^{-43}$	yes	167	47.9	5	10.00
830	SR03056	Proteasome regulatory particle 5	C. briggsae	CAP25033	$3e^{-41}$	no	175	21.1	2	6.86
831	SR00772	Prolyl carboxy peptidase like family member	C. elegans	NP_501598	$3e^{-33}$	yes	188	29.8	3	6.04
832	SS00857	ube1-prov protein	B. malayi	XP_00190173	$6e^{-50}$	no	220	8.2	2	5.70
833	SS01061	Proteasome regulatory particle	C. elegans	AAC19196	$4e^{-73}$	no	179	32.4	2	4.00
834	SR00536	Putative serine protease	C. elegans	P34528	$4e^{-50}$	yes	212	15.6	2	4.00
835	SS00471	Proteasome regulatory particle	B. malayi	XP_00189695	$5e^{-57}$	no	186	18.3	2	5.52
836	SR00789	26S proteasome non-ATPase regulatory subunit	B. malayi	XP_00189710	$3e^{-92}$	no	356	7.0	2	4.20
837	SR01639	Proteasome regulatory particle	C. briggsae	CAP27340	$4e^{-71}$	no	159	16.4	2	4.77
838	SS01358	Proteasome regulatory particle 9	C. briggsae	CAP23669	$3e^{-36}$	no	394	20.3	5	11.98
839	SR00582	26S proteasome non-ATPase regulatory subunit 3	B. malayi	XP_00189692	$5e^{-41}$	no	248	9.3	2	4.03
840	SR04945	26S proteasome non-ATPase regulatory subunit 7	B. malayi	XP_00189859	$2e^{-61}$	no	192	13.5	2	4.07
841	SR01336	CNDP dipeptidase 2	G. gallus	NP_00100685	$9e^{-59}$	no	181	15.5	2	5.30

Table 12a continued

	Cluster	BLAST Alignment	Species	Accession Number	E	SP	EST Lgt.	% Cov.	№ Pep.	UPS
842	SR00791	Proteasome regulatory particle 8	C. elegans	NP_491319	$8e^{-55}$	no	181	16.6	2	4.21
843	SR00261	26S proteasome non-ATPase regulatory subunit	B. malayi	XP_0019004 12	$2e^{-26}$	no	146	15.1	2	4.86
844	SR00895	Cysteine protease	B. malayi	NP_0011231 13	$3e^{-56}$	yes	180	13.9	2	4.15
845	SS01180	Capping protein beta subunit	X. laevis	ABL63902	$8e^{-63}$	no	192	19.8	2	4.00
846	SS00120	Chaperonin containing TCP-1	C. briggsae	CAP21934	$1e^{-73}$	no	199	16.6	3	6.00
847	SR01473	Proteasome beta subunit 1	C. briggsae	CAP32364	$1e^{-52}$	no	172	20.3	2	5.52
Structural proteins										
848	SS00087	Filarial antigen	B. malayi	AAB35044	$2e^{-47}$	trun	167	29.3	3	6.03
849	SS01134	Clathrin heavy chain	B. malayi	XP_0018957 63	$3e^{-81}$	no	197	18.8	2	4.18
850	SR02243	Filarial antigen	B. malayi	AAB35044	$8e^{-39}$	trun	188	13.8	2	4.04
851	SS00446	Myosin heavy chain B	B. malayi	XP_0018942 77	$7e^{-32}$	no	189	21.2	2	5.71
852	SR00642	Spectrin 1	C. briggsae	CAP32768	$4e^{-118}$	no	289	26.6	5	11.55
853	SR05253	Tubulin alpha chain	B. malayi	XP_0018991 89	$8e^{-88}$	no	187	16.0	2	4.66
854	SS03422	Coronin 1	C. briggsae	CAP20555	$8e^{-84}$	no	185	22.7	3	7.70
855	SR00651	Regulator of microtubule dynamics	C. elegans	NP_741608	$4e^{-35}$	no	166	54.8	7	17.33
856	SR00723	Kinesin like protein	C. elegans	BAA92262	$1e^{-22}$	yes	112	25.0	2	4.01
Developmental processes										
857	SR02703	Ribonucleoside-diphosphate reductase	G. gallus	NP_0010260 08	$6e^{-64}$	no	178	15.2	2	4.01
858	SS00301	ATP-dependent helicase	B. malayi	XP_0018943 68	$2e^{-95}$	no	204	9.3	2	5.73
859	SR01998	ATP-dependent RNA helicase p47 homolog	B. malayi	XP_0018931 12	$3e^{-85}$	no	188	13.3	2	5.52
860	SR04211	LEThal family member (let-767)	C. elegans	NP_498386	$2e^{-74}$	yes	241	26.1	5	14.29
861	SR03300	Prohibitin	A. elegantissima	ABD04181	$8e^{-10}$	trun	53	45.3	2	7.40
862	SS00996	Thioredoxin-like protein DPY-11	C. briggsae	CAP36281	$7e^{-69}$	yes	181	29.8	4	8.04
863	SS03435	LEThal family member (let-70)	C. elegans	NP_502065	$4e^{-79}$	no	147	36.1	3	6.00
Heat-shock proteins										
864	SR04512	Heat-shock protein 90	B. malayi	XP_0018954 98	$1e^{-59}$	no	189	12.2	2	4.72
865	SR02596	Heat-shock protein 40	S. feltiae	AAM81355	$5e^{-34}$	no	180	14.4	2	4.01
Cytoplasmatic										
866	SR04665	Ran binding protein 7	D. melanogaster	AAF68970	$5e^{-28}$	no	186	18.8	2	4.00

Table 12a continued

	Cluster	BLAST Alignment	Species	Accession Number	E	SP	EST Lgt.	% Cov.	№ Pep.	UPS
867	SS01442	KH domain containing protein	B. malayi	XP_001892806	$1e^{-153}$	no	600	6.2	3	6.25
868	SR03141	L-plastin	B. malayi	XP_001899119	$8e^{-43}$	no	168	56.5	6	12.01
869	SR00843	T complex protein 1, zeta subunit	B. malayi	XP_001891633	$1e^{-165}$	no	393	24.9	7	16.46
870	SS00427	T complex protein 1, gamma subunit	B. malayi	XP_001893392	$9e^{-42}$	no	155	25.8	3	6.00
871	SR02673	T complex protein 1, gamma subunit	B. malayi	XP_001893392	$9e^{-76}$	no	192	12.0	2	4.00
872	SR00184	T complex protein 1, delta subunit	B. malayi	XP_001895036	$2e^{-132}$	no	304	16.1	4	8.01
873	SR03920	N-myristoyl-transferase	G. gallus	XP_418632	$8e^{-78}$	no	199	12.1	2	6.58
874	SR00186	Importin alpha	G. gallus	NP_501227	$1e^{-91}$	no	267	18.4	3	6.00
875	SS00691	Type 2A protein phosphatase	B. malayi	XP_001898256	$1e^{-59}$	no	155	19.4	2	4.77
876	SS00827	Inositol-3-phosphate synthase	C. elegans	1VKO_A	$3e^{-69}$	no	171	18.7	2	4.69
Nucleic acid metabolism										
877	SR02620	28S ribosomal protein S15	C. elegans	Q9NAP9	$9e^{-43}$	no	147	28.6	3	6.06
878	SR03304	50S ribosomal protein L20	B. malayi	XP_001898530	$2e^{-45}$	no	176	16.5	2	4.42
879	SR00953	60S ribosomal protein L37a	B. malayi	XP_001902009	$2e^{-38}$	no	91	27.5	2	6.52
Other functions										
880	SR05116	Coatomer gamma subunit	B. malayi	XP_001893606	$1e^{-80}$	no	187	27.3	3	7.70
881	SR05228	Coatomer gamma subunit	C. elegans	Q22498	$3e^{-63}$	no	154	19.5	2	4.34
882	SR02919	Vacuolar H ATPase	C. elegans	NP_501399	$9e^{-52}$	yes	189	46.0	7	14.00
883	SR02459	Vacuolar H ATPase	C. briggsae	CAP23708	$2e^{-25}$	no	163	19.6	2	4.58
884	SS02557	Coatomer subunit beta	C. elegans	Q20168	$1e^{-48}$	no	159	22.6	2	4.00
885	SR02307	Human spastic paraplegia protein 7	C. briggsae	CAP33351	$5e^{-82}$	no	190	28.4	3	5.18
886	SR01070	ABC transporter class f	C. elegans	NP_498339	$2e^{-68}$	no	197	11.7	2	4.00
887	SS00943	5' nucleotidase family	B. malayi	XP_001898950	$2e^{-67}$	no	210	21.9	3	6.01
888	SS01963	DEAD/DEAH box helicase	B. malayi	XP_001894585	$3e^{-52}$	no	164	18.9	2	4.00
889	SR02300	Putative acid phosphatase	C. elegans	Q10944	$2e^{-37}$	yes	164	20.1	2	4.16
890	SR00549	Eukaryotic peptide chain release factor	B. malayi	XP_001902081	$3e^{-92}$	no	176	27.8	3	6.80
891	SS00163	Coatomer beta subunit	B. malayi	XP_001896419	$1e^{-68}$	no	160	16.3	2	6.41
892	SR00151	NOL5a	B. malayi	XP_001900200	$1e^{-45}$	no	140	26.4	3	6.05
893	SR02403	Interferon-induced protein kinase inhibitor	B. malayi	XP_001898046	$1e^{-45}$	no	189	19.6	2	5.23

Table 12a continued

	Cluster	BLAST Alignment	Species	Accession Number	E	SP	EST Lgt.	% Cov.	№ Pep.	UPS
894	SR03283	PCI domain containing protein	B. malayi	XP_001894001	$4e^{-19}$	no	181	17.1	2	4.00
895	SS01914	Sec23/Sec24 trunk domain containing protein	B. malayi	XP_00189388	$4e^{-54}$	no	174	13.2	2	5.64
896	SR01630	Extracellular solute-binding protein	C. beijerinckii	YP_001309625	0.71	yes	124	29.8	3	7.54
897	SR01269	COP9/Signalosome and eIF3 complex	C. elegans	NP_500618	$5e^{-13}$	no	198	15.2	2	6.94
898	SR00681	Succinyl-CoA ligase	C. elegans	P53588	$2e^{-60}$	no	183	16.9	3	6.00
899	SS01943	Nucleolar protein	C. elegans	NP_491134	$9e^{-64}$	no	211	37.9	5	10.77
900	SS00696	Sec61 alpha subunit	B. malayi	XP_00189970	$1e^{-95}$	no	215	10.7	2	6.50
901	SS01032	Sec61 alpha subunit	B. malayi	XP_00189970	$3e^{-77}$	no	167	17.4	3	8.54
902	SR02744	Arginine kinase	T. canis	ABK76312	$1e^{-77}$	yes	235	13.6	2	4.10
903	SR02742	Gut esterase 1	C. briggsae	Q04456	$5e^{-06}$	yes	172	20.3	3	6.00
904	SR01922	Arginine kinase	B. xylophilus	ACF74766	$1e^{-65}$	no	167	28.7	3	6.05
905	SR01851	Arginine N-methyltransferase	A. pisum	XP_00194903	$7e^{-65}$	no	194	16.5	2	4.29
906	SR00421	Arginine kinase	B. xylophilus	ACF74767	$3e^{-60}$	yes	171	17.0	2	4.00
907	SR00925	Mitochondrial carrier homolog	B. malayi	XP_00190043	$2e^{-21}$	no	172	46.5	6	14.66
908	SR00398	Membrane import protein	B. malayi	XP_00190010	$5e^{-59}$	no	172	41.9	4	9.52
909	SR00162	NUDIX hydrolase	P. marneffei	EEA20902	$8e^{-10}$	no	152	33.6	4	8.09
910	SS02146	Catalytic subunit of human protein kinase Ck2	H. sapiens	1NA7_A	$1e^{-50}$	no	140	32.1	3	7.70
911	SR01364	Chaperonin containing TCP-1 family member	C. elegans	NP_741117	$2e^{-105}$	no	250	9.6	2	4.17
912	SR03192	NUDIX hydrolase 7	C. elegans	NP_491336	$1e^{-10}$	no	135	31.1	3	9.40
913	SS01133	Sideroflexin	C. briggsae	CAP34984	$3e^{-59}$	no	208	14.4	2	4.00
914	SR00892	Phosphate carrier protein	C. elegans	P40614	$3e^{-96}$	no	229	23.6	3	6.70
915	SR00673	CED-3 protease suppressor	C. elegans	NP_491371	$4e^{-51}$	no	174	30.5	3	6.00
916	SS01099	Intracellular lectin	C. briggsae	CAP24562	$3e^{-101}$	yes	343	7.6	2	4.00
917	SR01496	Brix domain containing protein 1	C. elegans	Q9N3F0	$3e^{-49}$	no	176	14.8	6	13.10
918	SS00436	5'-methylthio-adenosine phosphorylase	D. melanogaster	Q9V813	$7e^{-51}$	no	157	13.4	2	4.00
919	SR01833	Haloacid dehalogenase-like hydrolase	B. malayi	XP_00189741 9	$3e^{-35}$	no	174	16.1	2	4.12
920	SR00919	Casein kinase 2, beta subunit	C. familiaris	XP_532075	$9e^{-80}$	yes	174	17.2	2	4.00
921	SS02677	Fibrillarin	C. briggsae	CAP30723	$3e^{-60}$	no	140	36.4	3	7.02
922	SS00225	Vacuolar H ATPase	C. briggsae	CAP29580	$3e^{-90}$	no	214	12.6	2	4.75
923	SR01806	RNA-dependent helicase	B. malayi	XP_00189368 7	$5e^{-71}$	no	196	17.9	2	4.30

Table 12a continued

	Cluster	BLAST Alignment	Species	Accession Number	E	SP	EST Lgt.	% Cov.	№ Pep.	UPS
924	SS00909	Glycoprotein 25L2 precursor	B. malayi	XP_001898370	$1e^{-94}$	yes	213	13.1	2	5.70
925	SR00672	Drosophila SQD (squid) protein family member	C. briggsae	CAP25946	$1e^{-70}$	no	299	22.7	5	10.01
926	SS01315	Ras-related protein Rab-7	B. malayi	XP_001894651	$2e^{-83}$	no	210	18.6	3	6.02
927	SR00174	MethylMalonic aciduria type B homolog	C. briggsae	CAP35686	$5e^{-23}$	no	126	22.2	2	4.86
928	SR01146	PQ loop repeat family protein	B. malayi	XP_001897577	$4e^{-66}$	no	236	9.3	2	4.21
929	SR00790	KH domain containing protein	B. malayi	XP_001897730	$2e^{-62}$	no	365	7.1	2	4.00
930	SR04120	GRIM-19 protein	B. malayi	XP_001895226	$2e^{-14}$	no	85	28.2	2	4.00
931	SR00515	TB2/DP1, HVA22 family protein	B. malayi	XP_001895412	$7e^{-34}$	no	170	18.8	3	6.00
932	SR00834	Adhesion regulating molecule	B. malayi	XP_001900662	$8e^{-48}$	no	182	23.6	3	9.24
933	SS01244	Outer-arm dynein light chain 3	B. malayi	XP_001901243	$2e^{-26}$	no	113	20.4	2	4.02
Not assigned										
934	SR02370	Hypothetical protein GA17399-PA	N. vitripennis	XP_001602595	$3e^{-06}$	no	139	21.6	2	4.00
935	SS02612	kap beta 3 protein	B. malayi	XP_001898232	$1e^{-21}$	no	178	28.7	4	9.70
936	SR02915	Hypothetical protein	B. malayi	XP_001893907	$2e^{-17}$	no	170	33.5	4	9.25
937	SR01344	kap beta 3 protein	B. malayi	XP_001898232	$1e^{-59}$	no	194	20.6	3	8.06
938	SR00342	KH domain containing protein	B. malayi	XP_001892806	$5e^{-37}$	no	193	28.0	3	7.23
939	SR05006	KH domain containing protein	B. malayi	XP_001892806	$4e^{-55}$	no	197	17.8	2	4.10
940	SR01472	Hypothetical protein CBG20766	C. briggsae	XP_001680038	$2e^{-06}$	no	172	20.3	3	6.02
941	SR02446	Hypothetical protein K10C2.1	C. elegans	NP_509079	$9e^{-18}$	yes	175	23.4	3	6.00
942	SR00198	Hypothetical protein C17G10.8	C. elegans	T34105	$1e^{-80}$	no	241	17.0	3	9.23
943	SR03213	Hypothetical protein Y110A7A.19	C. elegans	NP_491536	$2e^{-41}$	no	175	44.0	5	11.00
944	SR00506	Hypothetical protein C17G10.8	C. elegans	T34105	$2e^{-32}$	no	162	24.7	2	4.00
945	SS01006	Hypothetical protein F22F7.1b	C. elegans	NP_872194	$4e^{-63}$	no	183	17.5	3	6.08
946	SR02098	Hypothetical protein GD22375	D. simulans	XP_002078722	$2e^{-57}$	no	172	40.7	5	11.53
947	SR02121	Hypothetical protein R08E5.3	B. malayi	XP_001898778	$7e^{-18}$	no	191	22.5	2	4.00
948	SR00852	Hypothetical protein CBG20301	C. briggsae	CAP37348	$5e^{-75}$	no	238	34.0	6	12.00

Table 12a continued

Cluster		BLAST Alignment	Species	Accession Number	E	SP	EST Lgt.	% Cov.	№ Pep.	UPS
949	SR03012	Hypothetical protein CBG17652	C. briggsae	CAP35291	$4e^{-34}$	no	166	20.5	2	5.54
950	SR05199	CGI-66 protein	O. tauri	CAL52898	0.24	no	161	16.8	2	4.95
951	SR02357	Hypothetical protein K11B4.1	C. elegans	NP_493580	$4e^{-14}$	no	182	18.7	2	4.01
952	SR02147	Hypothetical protein CBG06032	C. briggsae	XP_00167031 1	$1e^{-61}$	no	211	27.0	3	8.63
953	SS00124	Hypothetical protein K06A5.6	C. elegans	NP_491859	$2e^{-100}$	yes	219	14.6	3	7.42
954	SR01337	Hypothetical 41.4 kDa Trp-Asp repeats containing protein	B. malayi	XP_00189867 2	$3e^{-44}$	no	176	22.2	2	4.44
955	SS03156	Hypothetical 41.4 kDa Trp-Asp repeats con-taining protein	B. malayi	XP_00189867 2	$1e^{-67}$	no	189	14.8	2	4.37
956	SR02500	Hypothetical protein C18B2.5a	C. elegans	NP_741750	$1e^{-09}$	no	166	17.5	2	4.26
957	SR00566	Hypothetical protein CBG02921	C. briggsae	XP_00167936 5	$3e^{-52}$	no	173	20.8	2	5.44
958	SR00297	Hypothetical protein CBG23813	C. briggsae	CAP20569	$1e^{-58}$	no	235	14.0	2	4.01
959	SR03077	Hypothetical protein C44B12.1	C. elegans	NP_500042	$1e^{-10}$	yes	191	16.8	2	4.01
960	SR03081	Hypothetical protein C44B12.1	C. elegans	NP_500042	0.001	yes	176	31.2	2	4.00
961	SR04360	Hypothetical protein F32B6.2	C. elegans	NP_501777	$7e^{-16}$	no	199	17.1	3	6.00
962	SR01450	Hypothetical protein CBG02952	C. briggsae	XP_00167936	$4e^{-07}$	no	144	22.2	2	4.02
963	SS01323	Hypothetical protein F47G9.1	C. elegans	NP_505879	$2e^{-81}$	yes	205	8.3	2	4.00
964	SR00159	Hypothetical protein 16129	D. ananas-sae	XP_00195567	$3e^{-43}$	no	159	12.6	2	4.00
965	SS01679	Hypothetical protein R06C1.4	C. elegans	NP_493029	$2e^{-24}$	no	87	67.8	3	6.01
966	SR00298	Hypothetical protein RUMTOR _02101	R. torques	ZP_01968524	4.5	no	139	19.4	2	4.00

9.1.23 Table 12b: Nematode RefSeq proteins found only in extracts from the free-living stages

	Acc. Number	BLAST Alignment	Species	SP	Frag. Lgt.	% Cov.	№ Pep.	UPS
967	gi|17540286	Hypothetical protein F38E11.5	C. elegans	no	1,000	4.4	4	8.18
968	gi|17554004	Hypothetical protein K02F3.2	C. elegans	no	702	5.6	3	7.16
969	gi|71994521	Hypothetical protein Y47G6A.10	C. elegans	no	782	6.3	3	7.03
970	gi|17537199	Hypothetical protein Y48B6A.12	C. elegans	no	620	7.4	3	6.01
971	gi|5973981	Sequence 33 from patent US 5827692	Unclassified	no	1,745	2.5	3	6.13
972	gi|62911496	COPII-coated vesicle component SEC-23	H. contortus	no	339	13.8	3	6.00
973	gi|3057042	P-glycoprotein	H. contortus	no	1,275	1.6	2	4.06
974	gi|17563250	Hypothetical protein F29G9.5	C. elegans	no	443	10.6	3	6.97
975	gi|17508701	Hypothetical protein F56H1.4	C. elegans	no	430	7.9	2	4.32
976	gi|17562296	Hypothetical protein K07C5.4	C. elegans	no	486	9.3	3	6.14
977	gi|32564660	Hypothetical protein T09B4.8	C. elegans	no	444	9.9	3	6.01
978	gi|71994394	Hypothetical protein T22D1.4	C. elegans	yes	586	4.4	2	4.08
979	gi|17554784	Proteasome regulatory particle	C. elegans	no	414	6.8	2	5.74
980	gi|17560378	Hypothetical protein F31D4.5	C. elegans	no	488	5.1	2	4.07
981	gi|71987372	Proteasome regulatory particle	C. elegans	no	398	16.3	4	12.20
982	gi|17555344	Hypothetical protein T28D6.6	C. elegans	no	366	13.7	4	8.77
983	gi|17533571	Hypothetical protein F29C12.4	C. elegans	no	750	4.9	3	6.74
984	gi|27803878	Cathepsin D-like aspartic protease	A. ceylanicum	yes	446	7.4	2	5.24
985	gi|17505779	Hypothetical protein C30F12.7	C. elegans	no	373	7.5	2	4.00
986	gi|17562816	Pyruvate carboxylase 1	C. elegans	no	1175	1.6	2	4.32
987	gi|17505290	Protein kinase 3	C. elegans	no	360	7.8	2	4.23
988	gi|665540	cAMP-dependent protein kinase catalytic subunit	A. caninum	no	360	7.5	2	4.13
989	gi|17532601	Chaperonin containing TCP-1 family member	C. elegans	no	549	3.8	2	6.74
990	gi|17554166	Hypothetical protein K11H3.3	C. elegans	no	312	6.7	2	5.70
991	gi|17509919	Hypothetical protein Y39G10AR.8	C. elegans	no	469	4.9	2	5.83
992	gi|71990788	NADH ubiquinone oxidoreductase	C. elegans	no	445	5.4	2	5.25
993	gi|17560974	Hypothetical protein F44G3.2	C. elegans	no	372	7.0	2	4.03

9.1.24 Table 13a: List of *Strongyloides* EST cluster numbers found in extracts from infective larvae and free-living stages

	Cluster	BLAST Alignment	Species	Accession Number	E	SP	EST Lgt.	% Cov.	№ Pep.	UPS
Oxydative metabolism										
994	SR00690	Epoxide hydrolase 1	R. norvegicus	EDL94842	$2e^{-43}$	no	177	24.3	3	6.00
995	SR01227	Isocitric dehydrogenase	C. elegans	Q93714	$1e^{-119}$	no	265	22.6	4	9.98
996	SS01954	Malate dehydrogenase 1	C. elegans	NP_498457	$1e^{-22}$	no	68	66.2	4	12.41
Carbohydrate metabolism										
997	SR01642	Oligosaccharyl transferase 48 kDa subunit	C. elegans	P45971	$2e^{-41}$	yes	169	36.1	4	8.01
Protein metabolism										
998	SR01770	Ribophorin II	B. malayi	XP_001897564	$1e^{-22}$	yes	198	41.4	6	12.43
999	SR00119	Initiation factor 1	C. elegans	NP_001022623	$2e^{-65}$	no	160	31.9	4	10.53
Protein digestion and folding										
1000	SS01434	Chaperonin TCP-1	C. elegans	NP_741117	$4e^{-127}$	no	304	19.1	3	6.00
1001	SS00930	Signal peptidase	B. malayi	XP_001894844	$9e^{-31}$	yes	131	29.0	3	6.33
Structural proteins										
1002	SR02148	Alpha-1 type IV collagen	C. elegans	AAB59179	$4e^{-65}$	trun	187	24.1	3	6.02
1003	SS00779	LEThal family member (let-2)	C. briggsae	CAP34098	$1e^{-86}$	yes	184	51.6	7	15.10
Developmental processes										
1004	SR00385	Mitochondrial prohibitin complex	C. briggsae	NP_490929	$2e^{-77}$	yes	173	30.1	4	8.50
Cytoplasmatic										
1005	SR00411	Serine/threonine protein phosphatase	B. malayi	XP_001894042	$2e^{-102}$	no	196	30.1	3	7.61
Other functions										
1006	SR03193	Conserved germline helicase 1	C. briggsae	CAP28848	$4e^{-94}$	no	204	41.7	7	17.53
1007	SR00697	Phosphate carrier protein	C. elegans	CAA53719	$3e^{-123}$	no	310	27.1	7	18.22
1008	SR00519	Vacuolar H ATPase 19	C. briggsae	CAP37981	$2e^{-04}$	yes	235	12.3	3	7.70
1009	SR00837	Cell death abnormality family member 11	C. briggsae	CAP27157	$9e^{-14}$	no	192	19.3	3	6.79
1010	SS01373	UNCoordinated family member (unc-108)	C. elegans	NP_491233	$4e^{-111}$	no	212	52.8	8	17.63
1011	SR00916	ATP synthase (atp-5)	C. elegans	NP_505829	$1e^{-63}$	no	189	40.7	5	10.75

Table 13a continued

Cluster	BLAST Alignment	Species	Accession Number	E	SP	EST Lgt.	% Cov.	№ Pep.	UPS
1012 SR00459	B-cell receptor-associated protein 31-like containing protein	B. malayi	XP_001897799	$6e^{-37}$	yes	180	26.7	4	8.02
1013 SS00082	Cell division cycle related family member 42	C. elegans	NP_495598	$3e^{-84}$	no	160	13.8	2	5.40
1014 SS01421	TspO/MBR family protein	B. malayi	XP_001898701	$5e^{-34}$	no	169	18.9	3	9.08
Not assigned									
1015 SS01735	Hypothetical protein C44B7.10	C. elegans	NP_495409	$2e^{-61}$	no	217	30.9	4	8.10
1016 SR02497	Hypothetical protein W01A11.1	C. elegans	NP_504650	$2e^{-50}$	yes	177	26.6	3	6.01
1017 SS02953	Hypothetical protein CBG02223	C. briggsae	XP_001668505	$4e^{-58}$	no	170	18.8	3	7.53
1018 SS00901	Hypothetical protein CBG03596	C. briggsae	CAP24461	$4e^{-27}$	no	126	19.0	2	4.59
1019 SR03116	Hypothetical protein CBG09819	C. briggsae	CAP29369	$9e^{-37}$	no	119	18.5	2	6.16

9.1.25 Table 13b: Nematode RefSeq proteins found only in extracts from infective larvae and free-living stages

	Acc. Number	BLAST Alignment	Species	SP	Frag. Lgt.	% Cov.	№ Pep.	UPS
1020	gi|17554158	Sarco-endoplasmic reticulum calcium ATPase	C. elegans	no	1,059	17.8	14	30.28
1021	gi|17555172	Clathrin heavy chain 1	C. elegans	no	1,681	3.9	4	8.76
1022	gi|71981301	Calpain 1	C. elegans	no	780	4.6	3	6.93
1023	gi|17551802	Hypothetical protein B0303.3	C. elegans	yes	448	9.2	3	6.00
1024	gi|2662171	Iron-sulfur subunit of succinate dehydrogenase	A. suum	no	282	12.1	2	4.02

9.1.26 Table 14a: List of *Strongyloides* EST cluster numbers found in extracts from parasitic females and free-living stages

Cluster	BLAST Alignment	Species	Accession Number	E	SP	EST Lgt.	% Cov.	№ Pep.	UPS
Oxydative metabolism									
1025 SR01689	Aldehyde dehydrogenase 1	*C. briggsae*	CAP39433	$3e^{-70}$	no	182	47.8	6	17.18
1026 SR01471	Glutaryl-CoA dehydrogenase	*C. elegans*	Q20772	$2e^{-48}$	no	147	25.2	2	5.28
1027 SR05222	Isovaleryl-CoA dehydrogenase	*C. elegans*	NP_500720	$8e^{-41}$	no	231	13.0	2	5.47
1028 SR00555	Trans-2-enoyl-CoA reductase 1	*C. elegans*	O45903	$4e^{-42}$	no	156	62.2	5	10.50
1029 SR01235	3-hydroxyisobutyrate dehydrogenase	*G. gallus*	NP_001006362	$2e^{-71}$	yes	241	45.2	6	12.01
Carbohydrate metabolism									
1030 SS02123	Glycosyl hydrolase 31	*B. malayi*	XP_001901753	$6e^{-38}$	yes	133	17.3	2	7.81
1031 SR01280	Transketolase	*N. vitripennis*	XP_001600105	$1e^{-54}$	no	168	58.9	6	9.70
Carbohydrate metabolism									
1032 SR01728	Acyl-CoA dehydrogenase	*B. malayi*	XP_001896268	$5e^{-62}$	no	227	38.8	4	9.89
Protein metabolism									
1033 SS01368	Polyadenylate-binding protein 1	*B. malayi*	XP_001894673	$1e^{-117}$	no	342	11.4	3	7.86
1034 SR01866	Calnexin 1	*C. briggsae*	CAP29597	$4e^{-66}$	yes	193	14.5	2	6.46
1035 SR00650	Protein kinase C substrate 80K-H	*B. malayi*	XP_001899873	$1e^{-44}$	no	179	25.1	3	7.78
1036 SR00911	Eukaryotic initiation factor 4A	*B. malayi*	XP_001894230	$4e^{-87}$	no	239	38.1	5	10.34
Protein digestion and folding									
1037 SR00683	T-complex protein 1 theta subunit	*B. malayi*	XP_001901094	$4e^{-81}$	no	257	19.5	4	8.06
1038 SR02145	T-complex pro-tein 1 eta subunit	*B. malayi*	XP_001896054	$4e^{-73}$	no	193	39.9	6	12.11
1039 SS00619	Hsp70/Hsp90-organizing protein	*A. pisum*	XP_001950745	$3e^{-31}$	no	121	25.6	2	6.38
1040 SR00372	Peptidase dimerisation domain containing protein	*B. malayi*	XP_001901221	$9e^{-43}$	no	180	12.8	2	4.84
1041 SR01035	Cathepsin Z-2	*B. malayi*	CAP26996	$5e^{-69}$	yes	206	10.2	2	4.70
1042 SS02652	Cathepsin Z-1	*C. briggsae*	CAP24890	$2e^{-77}$	yes	191	52.4	5	10.00
Heat-shock proteins									
1043 SR00659	HSP-6	*C. briggsae*	CAP28585	$1e^{-100}$	no	250	30.8	4	11.15
1044 SR05108	HSP-4	*C. briggsae*	CAP31983	$3e^{-92}$	yes	185	42.2	5	9.21
Cytoplasmatic									
1045 SS02061	Glycyl tRNA Synthetase 1	*C. elegans*	NP_498093	$2e^{-82}$	no	189	15.3	2	4.00
1046 SR04392	Phenylalanyl tRNA synthetase family member 1	*C. briggsae*	CAP31525	$1e^{-45}$	no	188	13.8	2	7.13

Table 14a continued

Cluster	BLAST Alignment	Species	Accession Number	E	SP	EST Lgt.	% Cov.	№ Pep.	UPS
1047 SR02406	Serine hydroxy-methyltransferase	C. briggsae	Q60V73	$1e^{-56}$	no	168	43.5	5	10.00
1048 SS02531	Cystathionine gamma-lyase	C. elegans	P55216	$6e^{-26}$	no	97	54.6	3	6.00
Nucleic acid metabolism									
1049 SR01034	60S ribosomal protein L3	T. canis	P49149	0.0	no	359	12.5	3	8.57
1050 SR01052	Ribosomal protein L7a	A. irradians	AAN05607	$4e^{-85}$	trun	242	24.0	5	18.36
1051 SR00994	60S ribosomal protein L7	B. malayi	XP_00189638	$1e^{-94}$	no	243	32.5	8	25.08
1052 SR01059	40S ribosomal protein S2	B. malayi	XP_00189924	$2e^{-101}$	no	271	24.4	5	11.97
1053 SR01033	Sr-rip-1	S. ratti	CAA05029	$1e^{-86}$	no	234	28.2	8	22.21
1054 SR02812	Ribosomal protein L10	B. malayi	XP_00189349	$4e^{-43}$	no	169	21.3	4	10.97
1055 SR01066	Large subunit ribosomal protein 16	K. sp. RS1982	ABR87201	$8e^{-75}$	trun	202	17.3	3	10.29
1056 SR00929	60S ribosomal protein L24	B. malayi	XP_00189277	$9e^{-50}$	no	158	13.3	2	5.60
1057 SS01467	60S ribosomal protein L23a	B. malayi	XP_00189243	$2e^{-46}$	no	151	18.5	2	5.76
1058 SR00936	Large subunit ribosomal protein 21	C. briggsae	CAP34592	$2e^{-56}$	no	160	16.9	3	9.27
1059 SR01219	Ribosomal protein L26	S. papillosus	ABK55148	$6e^{-52}$	trun	131	22.9	2	7.33
1060 SS01481	40S ribosomal protein S13	B. pahangi	P62299	$3e^{-72}$	no	151	23.8	3	8.06
1061 SR01012	40S ribosomal protein RPS15	F. foliacea	ABX44787	$4e^{-57}$	no	144	20.8	4	8.01
1062 SR00956	Ribosomal protein L27a	S. ratti	ABF69529	$2e^{-62}$	trun	152	23.0	3	6.85
1063 SR00968	40S ribosomal protein S24	B. malayi	XP_00189240	$6e^{-38}$	no	134	17.9	2	4.20
1064 SS03248	60S ribosome subunit bio-genesis protein	A. thaliana	NP_193312	$7e^{-20}$	no	89	30.3	2	4.00
1065 SS01399	Ribosomal protein L27	S. papillosus	ABK55149	$3e^{-61}$	trun	136	19.9	2	12.08
1066 SR02823	60S ribosomal protein L32	C. quinquefasciatus	XP_00184289	$8e^{-40}$	no	112	35.7	5	10.59
1067 SR01017	40S ribosomal protein S19S	A. suum	P39698	$2e^{-46}$	no	145	17.9	2	8.18
1068 SR03284	60S ribosomal protein RPL31	C. intestinalis	XP_00212890	$9e^{-42}$	no	120	18.3	2	5.64
1069 SS01488	40S ribosomal protein S12	B. malayi	XP_00190253	$4e^{-43}$	trun	129	42.6	4	8.00
1070 SR00938	60S ribosomal protein L35a	B. malayi	XP_00190205	$4e^{-33}$	no	108	24.1	3	7.27
1071 SS01295	60S ribosomal protein L38	S. cephaloptera	CAL69088	$3e^{-19}$	no	70	35.7	2	4.29
Other functions									
1072 SS01334	Protein phosphatase 2C	B. malayi	XP_00189708	$6e^{-109}$	no	350	7.1	2	5.52

Table 14a continued

Cluster	BLAST Alignment	Species	Accession Number	E	SP	EST Lgt.	% Cov.	№ Pep.	UPS
1073 SR05001	Methionine aminopeptidase 2	C. briggsae	CAP32448	$1e^{-87}$	no	198	19.2	3	6.04
1074 SR03831	La domain containing protein	B. malyai	XP_001895893	$8e^{-23}$	no	159	35.2	4	9.30
1075 SR01881	O-methyltransferase	B. malayi	XP_001897030	$1e^{-54}$	no	181	26.0	3	7.30
1076 SR02999	Coiled-coil protein	B. malayi	XP_001900931	$2e^{-22}$	no	155	43.2	5	11.58
Not assigned									
1077 SR02193	Hypothetical protein F10G7.2	B. malayi	XP_001893632	$2e^{-25}$	no	187	11.2	2	6.82
1078 SR00693	Hypothetical protein CBG00964	C. briggsae	CAP22287	$4e^{-09}$	yes	219	22.8	3	11.28

9.1.27 Table 14b: Nematode RefSeq proteins found in extracts from parasitic females and free-living stages

Acc. Number	BLAST Alignment	Species	SP	Frag. Lgt.	% Cov.	№ Pep.	UPS
1079 gi|32565889	UNCoordinated family member (unc-22)	C. elegans	no	7,158	0.5	3	6.57
1080 gi|71984108	Dihydropyrimidine dehydrogenase 1	C. elegans	no	1,059	9.5	8	16.13
1081 gi|17538396	Isovaleryl-CoA dehydrogenase	C. elegans	no	419	8.6	3	6.02

9.2 Galectin sequences

9.2.1 *Sr*-Gal-1

```
atggctgatgaaaaaaaagttacccagtaccatacaaatctcaacttcaagagaaattt
 M   A   D   E   K   K   S   Y   P   V   P   Y   K   S   Q   L   Q   E   K   F
gaaccaggacaaactcttattgtcaagggatctactattgaggaatctcaaagatttact
 E   P   G   Q   T   L   I   V   K   G   S   T   I   E   E   S   Q   R   F   T
gtaaatcttcattgtaaatctgctgacttctctggaaatgatgttccacttcatatttct
 V   N   L   H   C   K   S   A   D   F   S   G   N   D   V   P   L   H   I   S
gttcgttttgatgaaggaaaaattgttcttaatacttttctccaatggagattgggtaaa
 V   R   F   D   E   G   K   I   V   L   N   T   F   S   N   G   D   W   G   K
gaagaaagaaaaagtaatccaattaagaagggtgaaccattcgacattagaattcgtgct
 E   E   R   K   S   N   P   I   K   K   G   E   P   F   D   I   R   I   R   A
catgatgatagattccaaattatgattgatcaaaaagaattcaaagattacgaacataga
 H   D   D   R   F   Q   I   M   I   D   Q   K   E   F   K   D   Y   E   H   R
cttccactttcatccattactcatttctctgttgatggagatatttatttgaatactatt
 L   P   L   S   S   I   T   H   F   S   V   D   G   D   I   Y   L   N   T   I
cattgggaggaaaatattatccagtaccatatgaaagtggaattgcttctggattccca
 H   W   G   G   K   Y   Y   P   V   P   Y   E   S   G   I   A   S   G   F   P
gtagaaaagtcacttcttatctacgctactcctgagaagaaggctaaacgttttaacatt
 V   E   K   S   L   L   I   Y   A   T   P   E   K   K   A   K   R   F   N   I
aaccttttgcgcaaaaatggagatattgcccttcattttaaccccgttttgatgaaaag
 N   L   L   R   K   N   G   D   I   A   L   H   F   N   P   R   F   D   E   K
gctgtcgttcgcaataaccttcaagctggagaatgggtaatgaggaacgtgaaggtaag
 A   V   V   R   N   N   L   Q   A   G   E   W   G   N   E   E   R   E   G   K
atgccatttgaaaagggagttggatttgatctcaaaattgtcaacgaacaattctctttc
 M   P   F   E   K   G   V   G   F   D   L   K   I   V   N   E   Q   F   S   F
caaatttatgttaatggaaaacgcttctgcacttttgctcatcgttgtgatccaaatgac
 Q   I   Y   V   N   G   K   R   F   C   T   F   A   H   R   C   D   P   N   D
atttctggtcttcaaattcaaggagacctcgaacttactggaattcaaattaattaa
 I   S   G   L   Q   I   Q   G   D   L   E   L   T   G   I   Q   I   N   -
```

9.2.2 Sr-Gal-2

```
atgactcaagaaggatcatatccagttccttatcgtactaagcttactgagccttttgaa
 M  T  Q  E  G  S  Y  P  V  P  Y  R  T  K  L  T  E  P  F  E
cctggtcaaacattgactgtcaaaggaaaaactgccgaagattctgttcgtttttcaatt
 P  G  Q  T  L  T  V  K  G  K  T  A  E  D  S  V  R  F  S  I
aatcttcatactgctgcagcagattttctggaaatgatattcctcttcatatttctatt
 N  L  H  T  A  A  A  D  F  S  G  N  D  I  P  L  H  I  S  I
cgttttgatgaaggaaaaattgtattaaatacaatgagtaaatcagaatggggaaaagaa
 R  F  D  E  G  K  I  V  L  N  T  M  S  K  S  E  W  G  K  E
gaaagaaaaggaaatccatttaaaaagggtgatgatattgatattagaattcgtgctcat
 E  R  K  G  N  P  F  K  K  G  D  D  I  D  I  R  I  R  A  H
gacaacaaatttaccatttagctgatcaaaagagcttaaagaatatgatcatcgtctt
 D  N  K  F  T  I  L  A  D  Q  K  E  L  K  E  Y  D  H  R  L
ccactttcaagtgttactcatatgtcaatagaaggtgatattcttattaccaatattcat
 P  L  S  S  V  T  H  M  S  I  E  G  D  I  L  I  T  N  I  H
tggggtggaaaatattatccaattccttatgaaagtggaattggtggtgaaggaatttca
 W  G  G  K  Y  Y  P  I  P  Y  E  S  G  I  G  G  E  G  I  S
gttggaaaatctcttttattaatggaatgcctgaaaaaaaaggtaaacgttttttacatt
 V  G  K  S  L  F  I  N  G  M  P  E  K  K  G  K  R  F  Y  I
aatcttttgaaaaagaatggagatattgctcttcattttaatccaagatttgatgagaaa
 N  L  L  K  K  N  G  D  I  A  L  H  F  N  P  R  F  D  E  K
gccgttgttcgcaactctcttcttggaggtgaatggggtaatgaagagagaggaaaa
 A  V  V  R  N  S  L  L  G  G  E  W  G  N  E  E  R  E  G  K
attgtctttgaaaaaggacatggttttgatcttaaaataacaaatgaggaatatggtttc
 I  V  F  E  K  G  H  G  F  D  L  K  I  T  N  E  E  Y  G  F
caagtctttgttaatgatgaacgtttctgtacctatgctcatagagttgatccaaatgag
 Q  V  F  V  N  D  E  R  F  C  T  Y  A  H  R  V  D  P  N  E
attaatggactccaaataggtggagatgttgaaattactggaatccaacttttgtaa
 I  N  G  L  Q  I  G  G  D  V  E  I  T  G  I  Q  L  L  -
```

9.2.3 Sr-Gal-3

```
atgtctactgaaactcatttaccagtaccatatcgttcaaaacttactgatccctttgaa
 M   S   T   E   T   H   L   P   V   P   Y   R   S   K   L   T   D   P   F   E
cctggtcaaacattaatggtcaaaggtaaaacaattccagaatcaaaacgttttagtatc
 P   G   Q   T   L   M   V   K   G   K   T   I   P   E   S   K   R   F   S   I
aattttcattctggttcaccagatcttgatggaggtgatattccattccacatttctatt
 N   F   H   S   G   S   P   D   L   D   G   G   D   I   P   F   H   I   S   I
cgttttgatgagggaaagtttgttttaatacattcaataagggagaatggatgaaggaa
 R   F   D   E   G   K   F   V   F   N   T   F   N   K   G   E   W   M   K   E
gagaggaaatctaatccatataaaaaaggaagtgatattgatattagaatacgtgcccat
 E   R   K   S   N   P   Y   K   K   G   S   D   I   D   I   R   I   R   A   H
gacaacagatttgtcatatatgctgatcaaaaagaaattcatgaatatgagcatcgtgta
 D   N   R   F   V   I   Y   A   D   Q   K   E   I   H   E   Y   E   H   R   V
ccactttctacaataactcatttctcagttgatggtgatttaattcttaatcaagttaca
 P   L   S   T   I   T   H   F   S   V   D   G   D   L   I   L   N   Q   V   T
tggggtggaaaatattatccagttccatatgaaagtgggataaccggagatgggttagtt
 W   G   G   K   Y   Y   P   V   P   Y   E   S   G   I   T   G   D   G   L   V
cctggaaaaagtcttattattcatggaataccagaaaaaaaaggcaaaagtttcacaata
 P   G   K   S   L   I   I   H   G   I   P   E   K   K   G   K   S   F   T   I
aatatcttgaatgaaggaggagatgtagttctcagtttaatagtaaaattggagataaa
 N   I   L   N   E   G   G   D   V   V   L   S   F   N   S   K   I   G   D   K
catatcgttcgtaacgctaaaattggtaatgaatggggtaatgaagaaaaagaaggaaaa
 H   I   V   R   N   A   K   I   G   N   E   W   G   N   E   E   K   E   G   K
agtcctcttcagaaaggagttggttttgacttagaaatcaagagtgaaccatattcattc
 S   P   L   Q   K   G   V   G   F   D   L   E   I   K   S   E   P   Y   S   F
caaattttcataaacaatcaccgttttgctacatttgcacacagaactaatccagaagga
 Q   I   F   I   N   N   H   R   F   A   T   F   A   H   R   T   N   P   E   G
ataaagggtcttcagatttgtggtgatgttgaaattactggaattcagttggtttaa
 I   K   G   L   Q   I   C   G   D   V   E   I   T   G   I   Q   L   V   -
```

9.2.4 *Sr*-Gal-5

```
tttttagtttttgtactaatagcattgattgctactacttctgtatttggaagttcagat
 F   L   V   F   V   L   I   A   L   I   A   T   T   S   V   F   G   S   S   D
ggaaataaagaaagagaatatagaaaatttattggtgaaagagaattgccagttccattt
 G   N   K   E   R   E   Y   R   K   F   I   G   E   R   E   L   P   V   P   F
aaaacaaaagttacagaggcattaaagacaggtcatacaattcatgttatcggaactatt
 K   T   K   V   T   E   A   L   K   T   G   H   T   I   H   V   I   G   T   I
tctgaaaaaccaaaaagaattgatttcaattttcataaaggagcaagtgatgatgctgat
 S   E   K   P   K   R   I   D   F   N   F   H   K   G   A   S   D   D   A   D
atgcctcttcatctttcaattcgctttgatgaaggacttttccatagtaaaattgtttat
 M   P   L   H   L   S   I   R   F   D   E   G   L   F   H   S   K   I   V   Y
aacatttatgaaaatggtaactggtctgaaactgaacaaagaattgcaaatcctttcaaa
 N   I   Y   E   N   G   N   W   S   E   T   E   Q   R   I   A   N   P   F   K
gctaactctgaatttgatttaagagttcgtatcactgatggtgaatttaaaatgtatgca
 A   N   S   E   F   D   L   R   V   R   I   T   D   G   E   F   K   M   Y   A
aatagaaaagaaattggtgtttttaaacaaagaacctcaattgatggaattgatcatatt
 N   R   K   E   I   G   V   F   K   Q   R   T   S   I   D   G   I   D   H   I
tcaattaaaggagatttaaaatcattaaatcttttaaatatggtggaattcttttt gaa
 S   I   K   G   D   L   K   S   L   N   L   F   K   Y   G   G   I   L   F   E
actccatatactgctcttgcaaatcttacaccaggaaaaagacttgatatatctgctatg
 T   P   Y   T   A   L   A   N   L   T   P   G   K   R   L   D   I   S   A   M
ccaagaggaaaacgtattgatattgatcttttacgtaaaaatggtgatactgcactccaa
 P   R   G   K   R   I   D   I   D   L   L   R   K   N   G   D   T   A   L   Q
gtctcaattagatatggtgaatcagctattgtaagaaattctaaaaccggagaagtttgg
 V   S   I   R   Y   G   E   S   A   I   V   R   N   S   K   T   G   E   V   W
ggacaagaagatagaagtggaaaattcccattaaataaaaatgaactttttgatgttaca
 G   Q   E   D   R   S   G   K   F   P   L   N   K   N   E   L   F   D   V   T
attattaatgaaagttggtcattccaactttcttt aatggaaaaagatttggaacattt
 I   I   N   E   S   W   S   F   Q   L   F   F   N   G   K   R   F   G   T   F
gctcatcgtggagatattaatgatgttaagaaatttagaaaattactggagatgttgatata
 A   H   R   G   D   I   N   D   V   K   N   L   E   I   T   G   D   V   D   I
cttacagttactattaacgatgtcatttcttcataa
 L   T   V   T   I   N   D   V   I   S   S   -
```

9.2.5 Lactose-agarose bead isolation: Identified galectin peptides

Peptides matching to *Sr*-Gal-2 and having a confidence score of 99 %

Conf %	Sequence	Modifications	Cleavages	ΔMass	Spectrum
99	AHDNKFTILADQK		missed K-F@5	0.07219082	1.1.1.3327.1
99	FDEGKIVLNTMSK	Oxidation(M)@11	missed K-I@5	0.76366866	1.1.1.3530.1
99	FYINLLKK		missed K-K@7	0.61810422	1.1.1.4300.1
99	GDIALHFNPR		cleaved N-G@N-term	0.81714964	1.1.1.3716.1
99	ITNEEYGFQVFVNDER	Deamidated(Q)@9; Deamidated(N)@13		0.11906728	1.1.1.4920.1
99	KNGDIALHFNPR	Asn->Pro@2	missed K-N@1	0.68932021	1.1.1.3737.1
99	LPLSSVTHMSIEG-DILITNIHWGGK	Oxidation(M)@9		0.49217701	1.1.1.5147.1
99	LTEPFEPGQTLTVK			1.14115345	1.1.1.4316.1
99	NGDIALHFNPR			0.03423196	1.1.1.3689.1
99	NSLLGGEWGNEER			0.61632544	1.1.1.4211.1
99	NSLLGGEWGNEEREGK		missed R-E@13	1.43728602	1.1.1.3852.1
99	SLFINGMPEKK	Deamidated(N)@5; Oxidation(M)@7	missed K-K@10	0.51727676	1.1.1.3458.1
99	TKLTEPFEPGQTLTVK		missed K-L@2	0.47849241	1.1.1.4059.1
99	TNEEYGFQVFVNDER	Deamidated(N)@2; Iodo(Y)@5	cleaved I-T@N-term	0.64781696	1.1.1.5115.1
99	YYPIPYESGIGGE-GISVGK	Pro->pyro-Glu(P)@5		0.4480179	1.1.1.4997.1

Die VDM Verlagsservicegesellschaft sucht für wissenschaftliche Verlage abgeschlossene und herausragende

Dissertationen, Habilitationen, Diplomarbeiten, Master Theses, Magisterarbeiten usw.

für die kostenlose Publikation als Fachbuch.

Sie verfügen über eine Arbeit, die hohen inhaltlichen und formalen Ansprüchen genügt, und haben Interesse an einer honorarvergüteten Publikation?

Dann senden Sie bitte erste Informationen über sich und Ihre Arbeit per Email an *info@vdm-vsg.de*.

Sie erhalten kurzfristig unser Feedback!

VDM Verlagsservicegesellschaft mbH
Dudweiler Landstr. 99 Telefon +49 681 3720 174
D - 66123 Saarbrücken Fax   +49 681 3720 1749
www.vdm-vsg.de

Die VDM Verlagsservicegesellschaft mbH vertritt

Printed by Books on Demand GmbH, Norderstedt / Germany